# The language of smell

# The language of smell

Robert Burton

Routledge & Kegan Paul
London, Henley and Boston

First published in 1976
by Routledge & Kegan Paul Ltd
39 Store Street,
London WC1E 7DD,
Broadway House,
Newtown Road,
Henley-on-Thames,
Oxon RG9 1EN and
9 Park Street,
Boston, Mass. 02108, USA
Set in Monophoto Optima
and printed in Great Britain by
BAS Printers Limited, Wallop, Hampshire

ISBN 0 7100 8429 3

# Contents

# The use
## of smell

The act of smelling takes place as we breathe. Air is drawn through the nose and molecules of odour are caught by membranes in the nostrils spread over tiny scroll-shaped bones. Nerve cells in the membrane register the odours and a message is passed to the brain. In Man, the membranes add up to about the size of two postage stamps, if laid out, compared with the pocket-handkerchief size of the membranes in a fair-sized dog. The difference in size of the smell membranes in dog and Man give an idea of the difference in their ability to smell odours. The size of the part of the brain that deals with the nervous messages coming from the nose is also related to the importance of smell to an animal. It is relatively large in sharks and hedgehogs that hunt primarily by smell and small in birds and in primates, animals that rely largely on sight. The toothed whales lack this part of the brain completely and their sense of smell is completely defunct.

Before proceeding further with a description of the sense of smell, there are some problems of names and meaning to be cleared up. Smell is the same as odour and scent. In everyday language smell has the suggestion of unpleasantness, odour is more or less neutral, while scent is used either to mean a perfume or the trail left by an animal. Technically the science of smell is osmics (from the Greek *osme* – a smell) and odorous substances are osmyls but these words are rarely used outside the most technical scientific literature. The act of smelling is olfaction from the

## The use of smell

Latin *olfacere* – to smell. The membrane in the nose is the olfactory membrane and the parts of the brain dealing with smell are the olfactory lobes.

Modern Man not only has a relatively small olfactory apparatus, but he also makes little use of his sense of smell. He is generally conscious of smells only when they intrude upon him and he is made aware more of those that can be classed easily as pleasant or unpleasant. They are usually the latter: if an object is said to smell the implication is that it is unpleasant. This may be due partly to many common smells being 'not nice' – the word 'smell' almost implies nastiness – but it is more likely that modern Man has become almost completely reliant on his senses of sight and hearing, although primitive Man seems to have used his sense of smell as an accessory to sight and hearing when searching for food. Such a faculty has not been entirely lost. The perfumier and the wine-taster show the capabilities of a delicate nose, and Fraser Darling in *A Herd of Red Deer* comments on how his ability to stalk deer was considerably impaired when suffering from a cold. Although he relied mainly on sight and hearing, the sense of smell was playing an important, if unappreciated, role.

However, recent researches are providing evidence that Man is making a more significant use of his sense of smell than had been supposed previously. We are now realising that odours appear to affect our behaviour, although we may not be conscious of either the odours or their effects. Nevertheless, this does not alter the fact that Man is not the best subject for a study of the mechanism and uses of the sense of smell. Indeed, investigation has tended to centre on those animals in which smell is clearly a leading sense, the insects, dogs and mice for example. Over the past three decades, and particularly within the last ten years, there has been an intensive study of the sense of smell. From being regarded as a rather mysterious sense it is now realised that olfaction is playing as important a role in the lives of individual and groups of animals as do sight and hearing, and in some animals far more so. The brake to this realisation had been that the human observer, living in a world primarily of sights and sounds, cannot easily appreciate that an animal is using odours for learning the nature of events and objects in its environment. Because we cannot detect most of these odours ourselves, we have to deduce their meaning to an animal by observing its behaviour. So far, we have only the most clumsy instruments for detecting, recording and analysing odours and the properties of odours cannot be defined or measured in physical units, as are light and sound.

Since early times Man knew that some animals possessed a finer sense of smell than himself. In his role of exploiter of the natural world he learnt to employ these animals' senses for his own ends. The animal used most is the dog, which was domesticated 10,000–15,000 years ago. The hounds are dogs used for flushing game. Some, such as greyhounds and Afghans, employ sight but the foxhounds, bloodhounds and beagles hunt by smell. Interpretation of a hound's behaviour is a high skill and reaches its peak in the professional huntsman. Much is also learnt about the prowess of the quarry. There are numerous anecdotes that not only attest to the fox's acute sense of smell but to its appreciation, supposedly, of its own smell. Foxes have been described as breaking their scent trail by running through water or among flocks of sheep and even riding on the backs of sheep. It is hard to believe that even the wily fox can be so acutely aware that his own scent trail will lead to his undoing. The greater likelihood is that some of the observations on which these stories are founded are due more to coincidence – if a fox is chased across country, it would be surprising if it did not cross water somewhere along the route, or run through a flock of sheep.

Day-to-day observations of the behaviour of hounds and foxes, or other animals, can be very misleading. Douglas Gordon, a naturalist writing forty years ago, did not believe that hounds used their sense of smell to follow foxes. He considered that scent, the trail left by an animal, was something quite different from smell, even though both are detected by the nose. He suggested that scent was 'a matter of sympathy – a form of magnetism'. It is not unusual for mysterious sixth senses to be invoked where an animal is detecting qualities that register on none of our senses. But in every case where it has been examined closely, a 'sixth sense' turns out to be no more than a highly developed form of one of the normal senses, or even an unsuspected combination of more than one sense. The explanation of the seemingly miraculous powers of homing by birds through celestial navigation is an example. The birds are using the sharpest of eyesight, not the coriolis force set up by the Earth's rotation or an unknown 'sense of direction'. Similarly, some apparently marvellous powers have turned out to be manifestations of an acute sense of smell. There is the story, for example, of the American Indian who could locate the bodies of drowned persons with uncanny skill. It transpired that he was making use of a snapping turtle, a species that feeds on flesh, as an aquatic bloodhound. The turtle was released on the end of a long line and when it stopped swimming, the Indian probed the bottom of the river for the missing corpse.

# The use of smell

Careful observation and deduction, followed by well-contrived experimentation, as now being used, are producing exciting results. Perhaps the most significant of these is to show that odours emitted by animals form a language as effective as any code of visual or acoustic signals. They transmit to its fellows information about an animal's mental or physical state. Jane van Lawick-Goodall in *Innocent Killers* has described the odour language of African wild dogs. She described how one dominant female in the pack advertised her condition wherever she went, so that she was closely followed by the dominant male. Whenever she marked, that is, deposited her urine, the male would rush over and mark in the same spot. Van Lawick-Goodall puts it that presumably he was attaching a little extra warning to her advertisement, announcing to any other prospective suitor who might pass that way that this female was being escorted and interference could lead to trouble. Not only did the male place his warning mark on the exact spot where the female had marked but frequently he marked at precisely the same moment. Van Lawick-Goodall points out that this is no easy matter with the two dogs standing side by side and almost touching. If the male tries to hit the same blades of grass as the female by raising a hind leg sideways at the same time the two dogs will knock into each other. So the male stood on his front feet, his hind legs up in the air, his body vertical and was able to spray the same blade of grass at the same moment as the female. 'Their perfumes were inexorably mixed and none who passed later could doubt but that the lady fully accepted the proximity of her suitor.'

It is clear from this account that the simple and easily observable act of depositing urine speaks volumes to other wild dogs. Scent carried in the urine tells another animal concrete facts about the depositor, as seen, for example, in one of the most familiar actions of the domestic dog. The deposition of scent-laden urine and the subsequent sniffing by other dogs is an everyday sight in towns and similar behaviour is used widely by animals in the wild but there it is much less obvious. This is one of the main reasons why the role of scent has so often been overlooked. The language of smell becomes apparent only through the use of experiments. Two simple experiments will serve as examples.

The deermice of America are very similar to the European wood-mouse, in appearance and habits. There are some fifty-five species ranging from Alaska southwards to Colombia. They are found in many kinds of country, from swamps to semi-deserts, but each species usually has only a restricted habitat. In some places several species of very similar

*Two dogs identify each other by scent as they meet (Photo by Jane Burton)*

deermice are found in overlapping habitats, yet they do not interbreed and there must be a means by which the species are segregated. The mechanism of segregation was demonstrated by a simple experiment using two very closely related species, the Rocky Mountain deermouse and the Florida deermouse.

A deermouse of one or other of the two species was put in a cage with two side compartments. Prior to this, one side compartment had contained a Rocky Mountain deermouse and the other had contained a Florida deermouse. The test deermouse was allowed to wander into the side compartments at will and the records showed that it preferred to spend its time in the compartment where a member of its own species had left its odour. Presumably this is how deermice of the same species are attracted to each other even when sharing their habitat with other species. On the other hand, the Florida deermouse was not so selective about which compartment it spent its time in as the Rocky Mountain species. This is explained by the existence of only the one species of deermouse in Florida, which eliminates the necessity for segregation.

A further example will show how a simple experiment demonstrates the effect of odours on the social life of animals that would have gone quite unappreciated from mere casual observations of their habits.

## The use of smell

*The armoured catfish grubs for food at the bottom of South American streams. The entrance and exit of the U-shaped nostrils can be seen as short projecting tubes in front of the eyes. The six barbels around the mouth are organs of taste (Photo by Jane Burton)*

The odours in this case are waterborne and the animals are catfish. Catfish are nocturnal, they have poor eyesight but very well-developed senses of taste and smell. They often form packed communities and they can be induced to live peacefully, even resting one on top of another, in an aquarium tank. Normally, however, they stake out territories and two catfish kept in the same tank will fight. But if water from a tank containing a peacefully co-existing group of catfishes is circulated for several days into a tank containing a pair of rivals, the latter gradually lose their aggression and behave like communal catfish. The inference from this is that the peaceful catfish are releasing an 'anti-aggression' substance into the water to calm their fellows.

The observations on wild dogs and the experiments with deermice and catfish demonstrate three uses of the language of smell but they tell us very little about the sense of smell and how it operates. The nature of odours, that is the vocabulary of the language of smell, is the province of the organic chemist. The use of the language is investigated by ethologists (the students of animal behaviour) and ecologists (the students of animal relationships and communities). Between the two, physiologists investigate the workings of the sensory apparatus – the nose – with which an animal detects odours and transmits information to the brain where its behaviour is controlled. It is in the area covered by chemists and physiologists that the central problem of the sense of smell lies. It is not known for certain what properties make a substance odiferous or what are the properties that make one substance smell different from another. It is not known how an odiferous substance stimulates the sense organ in the nose. Until these problems are resolved smell remains a mystery sense. With the senses of vision and hearing there is no such problem. The colour of a light is related to its wavelength, as is the pitch of a sound, but there is no known spectrum for neatly defining odours or explaining their existence as there is with the wave theories of light and sound.

Current ideas on odour and olfaction are presented in Chapter 2. At the moment it is sufficient to know that an odour, or a smell or a scent, consists of molecules of a volatile substance that are carried in the air or, for aquatic animals, are soluble and carried in water. The sense of smell is, therefore, the detection of chemicals. In the language of smell, odours are secreted by one animal and carried in air or water to the olfactory organs of another animal of the same species. The odours act as messengers telling an animal about its fellows, as in the examples of wild dogs, catfish and deermice. They are called pheromones (from the Greek *pherein* – carry and *horman* – excite) and they are analogous to hormones, the chemical messengers secreted by an internal gland and carried by the blood to affect distant parts of the body. The standard definition of a pheromone is that it is a substance which is secreted to the outside of the body and received by a second individual of the same species, in which it releases a specific reaction, for example, a definite behaviour or a developmental process.

Pheromones may be secreted in the urine, as in wild dogs; in faeces, as in the hippopotamus which scatters its dung with its tail; from the mouth, as in food-sharing by ants and bees; or they may be secreted from special glands. Deer have pheromone-secreting glands on their faces and hoofs and the queen bee secretes 'queen substance' from her mandibular glands. There are two kinds of pheromone. Releaser pheromones cause overt changes in an animal's behaviour. If a minnow is injured it releases a pheromone that sends other minnows fleeing from the scene. The formic acid discharged by wood ants when their nest is disturbed brings other wood ants to their aid. Primer pheromones have a general effect, altering the physiological state of other animals rather than changing their immediate behaviour. Isolated female mice come into breeding condition when urine from male mice, containing a primer pheromone, is sprayed into their cages and queen substance prevents the development of rival queens in the beehive by suppressing their development.

The sense of smell confers several advantages on an animal. Its useful properties may be summarised as simplicity, specificity and spread. The sense of smell is simple because it does not involve the use of elaborate sense organs. Even in vertebrates the nose is far less complex than the eye and ear while, in the invertebrates, the olfactory organ may be little more than a few sensory cells linked to the central nervous system. The language of smell is also very simple. A single chemical can trigger a set piece of behaviour. Thus, the pheromone liberated by a female silkmoth is alone sufficient to initiate the sexual behaviour of the male. On detecting the pheromone, the male silkmoth immediately flies upwind until it finds the female and mates with her. This pheromone is also specific. Only male silkmoths react to the pheromone of the female silkmoth. Specificity is linked with simplicity in insects where a single chemical initiates mating behaviour. Simplicity is enhanced by the ability of a single pheromone to convey different messages in different contexts. Queen substance of honey-bees controls the development of other queens in the nest and also attracts drones during the nuptial flight.

The third advantage of smell lies in the propagation of odours by diffusion through, or movement of, the medium. Vision is limited because light travels in straight lines and an animal can receive visual communications only if it is looking at their source. Sound can travel round corners and reach a hidden receiver but sounds cease as soon as transmission stops. Odours are useful in circumstances where messages need to travel out of sight and need to last for extended periods of time. The latter is of particular importance as an odour can be deposited as a beacon continually emitting its message while the animal goes about its other activities.

These advantages can be seen in the uses of the language of smell summarised below, but they also have some disadvantages. Although the message of a pheromone may be specific it is not always private. Ambrosia beetles of the genus *Ips* are among the many beetles that bore tunnels under the bark of trees. The sexes are attracted to each other by pheromones liberated through the entrances of the tunnels. The disadvantage is that the pheromone also attracts predatory beetles to aggregations of *Ips* beetles. Another disadvantage is that odours are at the mercy of the medium that conveys them. It is common knowledge that game must be stalked from downwind so that the quarry cannot smell the hunter but the nature of the atmosphere is also important. Fraser Darling found that red deer are more readily stalked on a cold, dry day because scent does not carry well in these conditions. Scent is also impeded by the fine droplets of mist but on a fairly humid day, particularly if the humidity is variable, scent carries well. Huntsmen have long been aware of the importance of dampness of ground and air on the scent trails of foxes.

The use of smell as a language falls into five broad categories. The messages are often simple but animals with a complex social organisation, notably some mammals and the social insects (bees, wasps and ants), have developed quite complex repertoires of odour language which are used in ordering the lives of groups of individuals.

Among the simplest uses of odour are those of prey-seeking and the avoidance of predators. In the basic form a predator is guided to its prey by its odour, while the prey animal takes avoiding action when it smells the predator. A corollary to the predator/prey context is the use of odour to deter predators, as with the evil effluvium of the skunk, and of the use of an alarm pheromone to warn members of the same species of danger, as in the minnow already mentioned. Another simple function of odour is to guide an animal home. This can be the simple use of an odour that guides a salmon on its way upriver to the native stream or the sophisticated trail-laying behaviour of ants which mark their path with a pheromone so that they and their companions can find their way back to a source of food as surely as Theseus followed the trail of twine through the labyrinth of the minotaur.

Odours are the equivalents in the lives of many mammals of the songs and calls of birds. They regulate the organisation of social groups by allowing recognition of individuals and an appreciation of their social status; they bring the sexes together and are used to mark out territories. The most complex and interesting use of pheromones is the control of reproductive behaviour, bringing together members of opposite sexes and ensuring that the meeting occurs when both are in breeding condition. Pheromones are also used in communication between mother and offspring, particularly as recognition signals. Among fur seals, for instance, the cow leaves her pup on the breeding beach and goes to sea to fish. At intervals she comes ashore to suckle but she has to find her pup among the crowd of thousands of pups that mill about in the

## The use of smell

*A cow recognises her calf by smelling it. For many mammals the scent of the offspring is used not only for identification by the mother but is also needed to promote maternal behaviour*

colony. To narrow the search, she comes ashore and makes her way to the place where she left her pup. Every now and then she calls loudly, and hungry pups in the neighbourhood call back and come towards her. The cow sniffs them but ignores or drives away all except her own, whose identification is confirmed by careful smelling.

The fur seal mother has used three senses to attain her goal. First vision and then hearing were used to bring two animals together but it was smell that provided the final precise sensory data for identification. This is a point worth bearing in mind. We might say that an animal finds its prey, its mate or its young by smell but other senses are usually involved in the search. Except at the lower end of the evolutionary scale, where special sense organs are few or feebly developed, an animal's behaviour is based on integrated information from many senses but at any time, one of these senses will be more important than the others. Although this book is concerned with animals' use of the sense of smell it must not be forgotten that they have other senses. What will be shown is that zoologists are now becoming very much aware of how animals are making use of this sense where, previously, we only knew, at the most, that they *did* make use of it. We are beginning to appreciate what a world of smells conveys to a dog or a honey-bee and even that this is not a world from which we are totally excluded.

# 2 The mechanism of smell

The organ of smell in vertebrate animals is the nose but it is not the only sense organ sensitive to chemical stimulation. The sense of taste forms a second chemical sense, centred in the tastebuds lying on the tongue and mouth lining and concerned with the sensation of chemicals in solution. There is a third chemical sense, known simply as the common chemical sense. The last is often overlooked but its signals are often felt. The common chemical sense warns of dangerous substances, the irritants such as acids, alkalis, tear gases and spices. It is one of the oldest senses and in the most primitive organisms is probably related to the sense of touch. Single-celled organisms such as bacteria and the protozoans *Paramecium* and *Amoeba* take avoiding action when they bump into a solid object or if they meet chemical irritants in the water. There is no defined sensory system, merely a general sensitivity and reaction to adverse conditions. These animals also show a positive reaction towards substances on which they feed. At the evolutionary stage reached by the earthworm and its relatives there is an organised nervous system with rudimentary sense organs and the beginnings of a brain. From the simple chemical sensitivity, the senses of taste and smell are beginning to develop, but the common chemical sense retains its function as an alarm system throughout the animal kingdom. But in the evolution of the vertebrates, or animals with backbones, there has been a general increase in the importance of smell. The parts of the brain dealing with smell have become relatively larger and, in mammals, these parts have developed into the seat of the intelligence that sets them above other animals.

In soft-skinned aquatic animals, including the invertebrates, fishes and amphibians, sensory cells of the common chemical sense are scattered over the surface of the body, ready to warn of contact with dangerous substances, but in the primarily land-living reptiles, birds and mammals the skin is covered with a horny layer of dead tissue which is continually worn away and replaced from underneath. The main function of this layer is to prevent the body from losing water and to guard the underlying tissues from rubbing, but it also acts as a barrier against irritants. The common chemical sense in Man is found only in the soft tissues of the mouth, nose, eyes, anus, reproductive orifices and in open wounds where the protective horny layer is absent. When an irritant touches one of these sensitive spots our initial reaction is the same as that of *Paramecium*. We recoil to avoid the stimulus, but then more elaborate protective reactions are brought to play to clear the tissues of irritant. Pepper causes a flow of mucus and sneezing, while ammonia causes a copious flow of tears.

The common chemical sense merely warns of danger but its two off-shoots, taste and smell, have been elaborated to give varied information. They can warn of danger but they also detect edible material and the sense of smell gathers information about the nature and state of other organisms. The two senses are not always easy to distinguish although in everyday language tastes and smells are easily separated. In land animals taste is strictly restricted to the perception of substances dissolved in water, while odours are airborne. Sugar is tasted on the tongue but does not have a smell because it is not volatile at ordinary temperatures. It can, however, be smelt if sugar solution is placed directly on the sensory tissues of the nose. For Man, there are only four basic tastes: sweet, sour, salt and bitter. Everything else that we regard as taste or flavour is odour communicated to the nose. When the nasal passages are blocked by a head cold food becomes 'tasteless'. All that can be discerned is whether food is palatable and this is the original function of taste in the lower animals.

The distinction between tastes and odours in vertebrate animals is confirmed by anatomical differences. In the nose, the sense cells are nerve cells that connect directly with the tissue of the brain. Tastebuds are composed of sense cells related to the other cells in the skin. They do not connect directly with the brain but pass on their 'messages' through linking nerve fibres. In the fishes, all chemical stimuli are caused by substances dissolved in water but their senses of taste and smell can

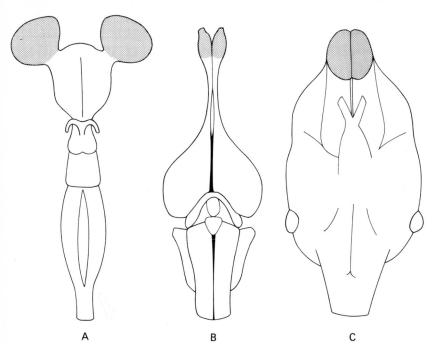

A          B          C

*Ventral view of the brains of (a) dogfish (b) alligator and (c) rat. The olfactory bulbs (stippled) are connected to the rest of the cerebrum by the paired olfactory tracts. The olfactory bulbs are relatively large in the dogfish and rat, both of which are reliant on the sense of smell.*

still be distinguished on this anatomical basis. The distinction between the two senses breaks down in the lower invertebrate animals and, to avoid argument or confusion, the all-embracing term chemoreception is used. This is a clumsy word and one can hardly talk of an animal chemo-receiving a substance but chemoreception must be used when discussing the lowest animals which have neither proper sense organs nor differentiation of chemical senses.

The restriction of taste to four qualities and of the common chemical sense to one quality render them useless as a means of communication. Only the sense of smell is organised on a level comparable with the senses of sight and hearing and is capable of giving an animal precise information about its surroundings. The nasal organs of fishes are located in paired nostrils in the forepart of the head. In bony fishes they lie on the side of the head; in sharks and their relatives they lie under the head. There is no connection with the gills, smelling and breathing being completely separate except in the coelacanth and the lungfishes where the nostrils open into the roof of the mouth. This is correlated with the development of air-breathing as it provides an air passage to the lungs via the mouth and throat. This type of fish gave rise to the amphibians and the rest of the land vertebrates in which there is a palate separating the nasal cavities from the mouth. Air is led from the nostrils through the nasal cavities to the throat, so by-passing the mouth and enabling the animal to breathe while chewing its food.

In Man and other mammals, the sensory part of the nose lies in the olfactory clefts, two cavities each lying above one of the main air passages of the nasal cavities. During normal breathing air passes to and fro in the nasal cavities with only slight eddies entering the olfactory clefts. Consequently only the strongest smells register on the sense organ. Detection of weaker smells is achieved by sniffing, to cause a mass of air to flow into the olfactory clefts. Inside each olfactory cleft the odorous molecules fall on a sheet of yellowish tissue, the olfactory epithelium or smell membrane, which is moistened with mucus. In Man, the two sheets contain about five million receptor cells where the physiological process of smelling takes place. In a dog, the area of tissue is 50 times as great and contains some 220 million receptor cells.

In many amphibians, reptiles and mammals there is an auxiliary sensory area called the vomeronasal organ or Jacobson's organ. It appears as a blind side passage in the nasal cavities of amphibians and most reptiles but in the lizards and snakes and in mammals it has no connection with

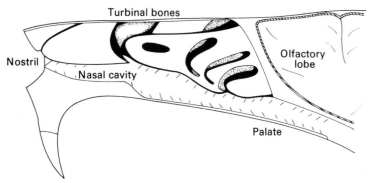

*Section through the head of a rat. Most of the muzzle is taken up by the scroll-shaped turbinal bones lying in the nasal clefts. The olfactory membrane is spread over the turbinals to receive odorous molecules carried in air circulating around them.*

the air passages but opens through a narrow duct into the roof of the mouth. The vomeronasal organ is missing in Man but it seems that other animals use it for smelling fluids in the mouth. In the forked-tongued lizards and snakes it has become a specialized organ of smell. The well-known flickering of the tongue is used to pick up scent particles and pass them to the vomeronasal organ.

Microscopic examination of the tissue in the olfactory clefts shows that there are four kinds of tissue cell. The sense cells, or receptors, form a layer of tissue called an epithelium in which each is surrounded by supporting cells that form a kind of packing, isolating the receptors from one another. Around the edges of the olfactory epithelium there are cells that secrete mucus and cells with cilia that continuously beat to propel the mucus over the olfactory epithelium. Each receptor terminates in a frond of extremely long, fine cilia that lie in the mucus stream. In dead tissues examined under the microscope these cilia look like wire wool but in life they stream out like pond-weed trailing in a stream. The

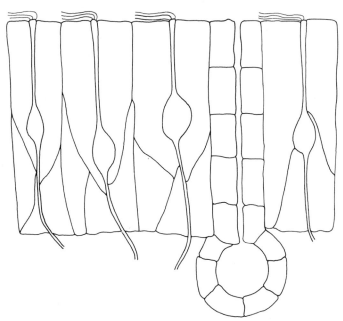

*Part of the olfactory epithelium showing the receptor cells with their terminal masses of hair-like cilia. They are surrounded by supporting cells and the surface of the epithelium is kept moist with secretions from the bulb-shaped nasal gland.*

cilia greatly increase the surface area of the olfactory epithelium and are the site of interaction between odorous molecules and the receptors. It appears that the odorous molecules brought into the nasal clefts dissolve in the mucus and somehow stimulate the receptors to create an electric charge, a process known as transduction. The charge initiates a nerve impulse that is transmitted to the olfactory lobes of the brain.

There are two main problems concerning this mechanism, neither of which has yet been satisfactorily resolved. There is the question of how the odorous molecule stimulates the receptor cell on which it impinges and, related to this, how molecules of various substances affect the receptors differentially to give the sensation of different odours. The two can be summarised as the problems of sensitivity and of discrimination. The great stumbling block is that there are thousands of known odours but there is no good classification that reduces them to order in the same way as sound and light are classified by wavelength. Several theories of odour classification have been proposed in order to give a platform for investigation of the transduction of odour molecules into nerve impulses but before these are discussed it would be best to set down some observed facts about the transduction mechanism.

Investigation of the olfactory epithelium by techniques that have been employed on receptors in the eyes and ears has not been easy because the olfactory receptors cannot be easily separated from the supporting cells (there are no supporting cells in the retina of the eye for example. Each receptor has a diameter of only ½ micron (1 micron = 1 thousandth of a millimetre), only a little larger than the limit of resolution by a light microscope. Furthermore, the sense of smell is extremely sensitive; even Man can detect contaminants in certain substances, such as menthol and saffrol, at much lower concentrations than can be detected by standard methods of the chemist. It seems that no more than eight molecules need to hit a receptor cell for a nerve impulse to be generated and impulses from forty receptors are needed for the smell to be perceived.

The working of smell receptors has been more easily studied in insects. Their sense organs are more accessible than those hidden deep in the nose of vertebrate animals and they are more simple. Insects are also useful for the investigation of sense organs because their behaviour is simple and stereotyped. A stimulus results in a fixed behavioural reaction which can be observed and measured. Thus the male silkmoth starts to search for a female when a pheromone, bombycol, from the female wafts over his antennae. It has been found that a male silkmoth will react when there are only one hundred molecules of bombycol per cubic centimetre of air. There are 25,000 receptors on each antenna and each one reacts with a nerve impulse when struck by one or two bombycol molecules. It then needs several hundred nerve impulses to start the silkmoth searching.

A possible way in which the bombycol molecule or any other odorous substance stimulates the receptor can be postulated from what is known of the mechanism of receptors in other sense organs. In its resting state a receptor or other cell in the nervous system is polarised, that is it maintains an electrostatic charge across its enclosing membrane, the inside being negative with respect to the outside. The charge is caused by a high concentration of charged potassium ions and a low concentration of sodium ions in the fluid inside the membrane, as compared with the concentrations in the fluid outside. When the cell is stimulated, sodium ions flow into the cell, the charge is reversed and a wave of depolarisation spreads along the surface of the membrane.

How the depolarisation starts is the central problem of nerve physiology. Somehow the character of the nerve membrane is altered to make it permeable to sodium ions. In the touch receptors that lie in the skin, mechanical deformation probably alters the properties of the membrane and allows the sodium ions through. The same will be true of the receptors in the ear that are stimulated by sound waves. In the nose, it is presumed that odorous molecules impinge on the receptor membranes and distort the molecules in the membrane so that the sodium ions can pass through. The problem then is to find how molecules of different substances can have different effects on the receptors. The solution has been sought by examining the properties of odorous molecules to see if molecules of substances having a similar odour have any chemical or physical feature in common.

Molecular structure was postulated as the basis of smell two thousand years ago. In 27 BC Lucretius wrote 'such substances as agreeably titillate the senses are composed of smooth, round atoms. Those that seem bitter and harsh are more tightly compacted of hooked particles and accordingly tear their way into our senses.' Lucretius, one feels, was thinking of smell as a delicate sense of touch on the molecular level. It is possible to determine accurately the chemical composition, shape and other properties of molecules, so how neatly do the odours of substances correlate with molecular structure? What is the olfactory code that gives a common factor to all substances with an odour of bitter almonds, such as hydrogen cyanide, nitrobenzene and benzaldehyde, yet will distinguish them from substances having, say, a camphoraceous smell? The supposition is that there are a number of receptor types, each of which is sensitive only to one group of similar odours. Like the three primary colours – red, green and blue – which produce other colours when mixed, primary odours are the units from which any of the thousands of known odours can be made. It remains to be decided which are the primary odours and which characters they have that could selectively stimulate their respective receptors.

In describing any biological mechanism for a non-specialist audience who may have no more than a passing interest in the subject, whether it be the sense of smell, the evolution of fishes or protective coloration, it is an advantage, to both reader and author, to be able to present a simple, clear picture. The workings of a motor-car engine can be described simply and confidently, starting with the theory of the four-stroke cycle of combustion and giving details of each mechanical part. The resulting description is beyond argument but in biology the story is never so simple. There are often gaps in our knowledge and some facts may be seemingly contradictory or open to varied interpretation. The result is that specialists working on the problem from different angles arrive at different conclusions. Yet each is, not surprisingly, likely to advertise his theory as the sole truth. For the scientist, conflicting theories add interest to the problem as each new piece of evidence that appears can be tested to see which theory it supports. Passions may well run high as the relative merits of theories are attacked or evidence is warped to fit existing ideas. But, eventually, the excitement will die down, as one theory gains majority support and goes into textbooks as the truth. It is as well to remember that there are many subjects which have gained general agreement and have been described confidently and dogmatically in textbooks but have later had to be completely rethought and rewritten.

The sense of smell is at the stage of debate. There have been many theories concerning the classification of odours and the way that they selectively stimulate the nose to give us particular sensations. Chemical composition is clearly ruled out as Table 1 shows that related odours can be chemically very different. Moreover, chemically similar compounds can have different odours. This has for long been the big stumbling block to any theory. At the moment there are two contenders which overcome this problem: the stereochemical theory of J. E. Amoore and the vibrational theory of R. H. Wright. Both proponents argue their own claims persuasively and have amassed such a wealth of support that, to read one exposition alone, one feels that the matter is virtually settled.

To read both leaves one in a quandary. But both must be presented.

TABLE 1 *Chemical composition of substances having camphoraceous odours*

| | |
|---|---|
| d-camphor | $C_{10}H_{16}O$ |
| hexachlorethane | $C_2Cl_2$ |
| trinitroacetonitrile | $C_2N_4O_6$ |
| silicononyl alcohol | $Si(C_2H_5)_3.C_2H_4OH$ |
| pentamethyl ethylalcohol | $C_2(CH_3)_5OH$ |
| benzene hexachloride | $C_6H_6Cl_6$ |
| durene | $C_6H_2(CH_3)_4$ |
| tetrachlornaphthalene | $C_{10}H_4Cl_4$ |

Amoore's stereochemical theory is a development of an earlier idea, and could even be thought to be the descendant of Lucretius' theorising. Amoore has pointed out that three of the compounds in Table 1, d-camphor, hexachlorethane and trinitroacetonitrile have very little in common except their stereochemical formulae, or molecular shape. Their molecules are basically spherical with the same diameter, as are over 100 substances with a camphoraceous odour that have been investigated and further work has shown more close correlations between other groups of similar odours and the size and shape of their molecules. Amoore concluded that these groups are the primary odours ( ? ) and that there are seven of them. In the simplest form of the theory a molecule of each primary odour is seen as fitting into a 'site' in the membrane of the receptor cell as a key fits a lock, as follows:

| Primary odour | Shape of molecule | Shape of site |
|---|---|---|
| camphoraceous | spherical | elliptical bowl |
| musky | disc | elliptical bowl but larger |
| floral | disc with tail | bowl with trough |
| pepperminty | wedge | V-shaped trough |
| ethereal | rod | trough |
| putrid | negatively charged | positive charge |
| pungent | positively charged | negative charge |

So, tissue in the nasal clefts contains seven kinds of receptors, each responding to the particular shape of the molecules of a primary odour. Secondary odours, made of molecules whose shape allows them to fit more than one kind of site, will stimulate more than one kind of site and form a composite odour, as the combination of the primary colours red and green gives the secondary colour yellow.

The seven primary odours were established from the subjective sensations reported by panels of human odour testers. It is very likely that there are other kinds of primary odour in the animal kingdom. It is reasonable to suspect that dogs have a larger repertoire of primary odours than we do and that some insects may react only to one odour – a food plant or the attractant from the opposite sex – and have only one kind of receptor. These receptors may be different from any of ours, as we cannot smell the sex attractants of some moths, for instance.

Wright, on the other hand, considers that seven receptor sites are too few to give the number of combinations required for all the odours that we can smell. He considers that there should be twenty to thirty kinds of receptors. In his vibrational theory he suggests that the odoriferous nature of a molecule is imparted by its low-frequency vibration. Molecules of different substances have characteristic, if complex, vibrations, each molecule vibrating at several frequencies. Wright suggests that in each kind of receptor there are molecules that vibrate at the same frequency as one of the primary odours. When an odour molecule meets the right receptor molecules the two resonate together at a slightly different frequency and cause a 'puncturing' and consequent depolarisation of the receptor membrane. This, as we have seen, is the start of a nerve impulse. Support for this scheme is given by analysis of the molecular vibrations of chemicals that attract the oriental fruit fly (one of the many species of *Drosophila* that are much used as research animals). Each chemical was found to have a common peak of vibrations at one particular frequency, whereas there was a conspicuous lack of vibrations at this frequency in chemicals that do not attract the fly.

Here we have two rival theories, both of which have good supporting evidence. According to approved scientific method, a good test of a theory is that it can be used to predict events or properties. If a search then reveals these events or properties, the theory is substantiated. Both theories of smell have been tested in this way. To test Amoore's theory, a number of new chemical compounds were synthesized and predictions of their odour were made from stereochemical measurements of their shapes. The shapes and odours were found to tally and so confirm Amoore's theory. Wright tested his theory by studying the odours that attract wasps. As with his work on fruit flies, these odours were found to have similar vibrations. He then predicted that certain other compounds ought to attract wasps because they had the right molecular vibrations. His predictions were substantiated when tested.

The drawback to both theories is that neither can be proved absolutely. Until it becomes possible to investigate the properties of a single receptor cell and actually find the receptor sites, the protagonists can do no more than gather overwhelming circumstantial evidence through observation

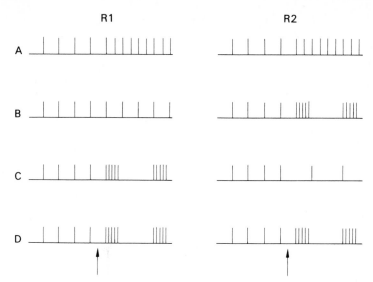

*Hypothetical response of two receptor cells, R1 and R2, to four odours, A, B, C and D. At rest, each receptor fires a slow, steady train of nerve impulses. On stimulation ( ↑ ), the train of impulses speeds up, slows or is emitted in bursts, according to the nature of the stimulus. R1 distinguishes odours A, B and C but cannot distinguish C and D. R2 can distinguish C from D but cannot separate B from D. Taken together, the responses of the two receptors allow the brain to distinguish all four odours.*

and prediction. What can be observed is the nerve impulses coming from the receptors. Ignoring how odours stimulate receptors, it has been found that a single odour, e.g. methanol, blown into the nose of a frog, causes some receptors to fire nerve impulses, others are inhibited and there is no effect on a third group. Even within the first group some receptors fire rapidly, some slowly and some in bursts. Each receptor is, therefore, capable of distinguishing a few odours or a group of odours and the brain recognises an odour from the reactions of several receptors, as shown above. A receptor emits a train of impulses even when there is no odour to stimulate it but, when stimulated the emission of impulses can speed up, slow down or be produced in bursts, or there may be no response.

The more receptors there are in the nose, the greater will be its powers of discrimination. The brain, therefore, does not receive a simple message in the form of a train of nerve impulses from receptors specific to each odour, but receives a very complex pattern of information, rather like a chord struck on a piano in which a selection of notes combines to form one sound. The brain recognises the pattern as an odour. A similar mechanism is thought to be present in the ear, where sounds of one pitch stimulate a number of receptors to produce a pattern of nerve impulses.

It is unsatisfactory to leave the mechanism of smell in an uncertain state and it might be thought that the latter part of this chapter would have been better left out to avoid confusion and despondency in the reader's mind. But it is not necessary to know how the nose works in order to appreciate the role played by scent in directing the behaviour of animals. Nor, to expand an earlier simile, is it necessary to know how a motor car engine works in order to study traffic problems. However, the performance of a car in traffic can be better appreciated with some idea of what is going on under the bonnet. So, too, is it with the study of smell and behaviour.

# 3 Attracting and repelling

One of the more exciting recent advances in zoology is the realisation that lowly animals that appear to be living entirely independent lives may, in fact, be in constant communication with each other. The means whereby this happens are chemical and, more importantly, through the use of pheromones. To understand this more completely we need to start with the lowest forms of life and trace the origins of the sense of smell and the development of its use in communication between animals. All living things must be able to appreciate the nature of their environment and to react appropriately to beneficial or adverse surroundings. For the lower forms of life, the environment is fluid and the chemical nature of this fluid is vital to the wellbeing of the organisms living in it. Certain substances in the fluid may be damaging while others are used as food. Among plants and those animals that have limited power of movement, like sea anemones, or are unable to move, like corals, reactions can only be passive. Food has to come to them and injurious substances have to be avoided by a physical barrier such as a tough skin. Where the organism can move freely, a sensing and guidance system is employed so that injurious conditions can be avoided and food actively sought.

A reaction to chemicals in the surrounding medium is seen in organisms as primitive as the bacteria. These are usually thought of as rather passive organisms, but some kinds are capable of swimming by the beating of cilia or flagella and can be attracted to concentrations of glucose and other sugars, glycerol and dissolved oxygen. If a bacterium happens to pass from a region of high concentration of an attractive solution to a lower concentration it immediately recoils and sets off in a new direction, either by turning or by simply reversing the beat of its cilia or flagella. The turns are repeated until the bacterium finds itself travelling back into the higher concentration.

Nothing that resembles a receptor or a nerve can be seen in a bacterium and nothing is known of the way its reaction to chemicals works, except that *Escherichia coli*, a bacterium that lives in the human digestive system, has a specific protein which binds with the sugar galactose and is in some way involved in stimulating the bacterium to move towards this sugar. If the mechanism could be found it would probably solve the problem of the function of olfactory receptors in higher animals. The cell wall of a bacterium may well react to a stimulating substance in the same way as does a receptor cell in the nose, becoming depolarised and initiating a wave of depolarisation over the surface membrane. Depolarisation in the nasal receptor results in a nerve impulse that travels to the brain, whereas in a bacterium the depolarisation spreads to the base of the flagella. Whatever the mechanism may prove to be, it is already known that a bacterium, such as *Escherichia coli*, must have at least five receptor systems for different chemicals.

On the next level higher than bacteria in the evolutionary scale are the single-celled protozoans. Most can move by means of cilia or flagella, or by pseudopodia as in *Amoeba*. Some contain chlorophyll and can form carbohydrates by photosynthesis; these are often classed as plants while the rest of the protozoans are regarded as animals. The distinction is rather arbitrary as both the plant and animal kingdoms arose from the Protozoa. Like bacteria, protozoans react to certain substances in their medium and some that feed by engulfing minute particles are able to distinguish between edible and non-edible material, accepting the former and rejecting the latter. How the chemical nature of the particles or solution is sensed and how the appropriate action is initiated is as much of a mystery in protozoans as in bacteria.

However, the first use of pheromones can be discerned in the protozoans. They reproduce by two methods: asexually, by splitting or budding, and sexually, by the fusion of two cells. In some kinds of protozoan both methods are used. Some species of *Chlamydomonas*, a protozoan containing chlorophyll that swims by means of two flagella, reproduce sexually through the fusion of two individuals, as well as asexually by splitting. In some instances, in order to mate, the two individuals have to come from separate clones. A clone is a population that has descended from a single individual by asexual reproduction and its members are, therefore, genetically identical. Fusion of individuals from different clones ensures a genetic mixing or interbreeding.

When two clones of these species of *Chlamydomonas* are mixed, they form clumps which later separate into fused pairs. The aggregation is caused by a pheromone secreted from the flagella and the pheromone from members of one clone causes clumping only in members of another clone. Thus, clumping only takes place between two different clones and interbreeding is assured.

Related to *Chlamydomonas* is a colonial organism called *Volvox*, which consists of hundreds of *Chlamydomonas*-like cells embedded in the surface of a ball of jelly. Along with other colonial protozoans, *Volvox* is thought to be among the forerunners of multicellular plants. It lives in pond water and, being half a millimetre across and bright green, can be seen with the naked eye. *Volvox* can reproduce asexually, by a single one of its cells repeatedly splitting to form a new daughter colony inside the parent colony, or else by sexual reproduction. In this, one cell forms a female macrogamete, or egg, which is fertilised by one of a number of microgametes, or sperms, formed by multiple division of another cell and liberated into the water by another *Volvox*. A pheromone secreted by a 'female' *Volvox* stimulates the asexual buds in nearby *Volvox* to develop into packets of male sperms instead of forming independent daughter colonies. Here is the first indication of chemical messages being used to alter the development of another organism. The *Volvox* pheromone is a primer pheromone whose function is to bring individuals into a suitable state for reproduction at the same time. Similar pheromones are to be found throughout the animal kingdom.

A particular advantage of the chemical sense to the protozoans and simple multicellular organisms is that complex sense organs are unnecessary for receiving chemical information. They can, however, react only generally to the stimuli of light and vibration and cannot resolve them into any form of visual or auditory image that might be used for communication. As a result, chemical sensation remains the most important sense among lower animals for both finding food, avoiding enemies and meeting members of the opposite sex. Eyes and ears only become important in the levels of organisation reached by the arthropods, including insects and spiders, the molluscs, such as squid, and the vertebrates.

## Attracting and repelling

It may be perhaps stretching a point to include the detection of odours for finding food or avoiding enemies in a discussion on the language of smell, but such odours tell one animal about another and are, hence, a form of communication. This is particularly so in the case of repellent odours used to drive off predators. Among the lower forms of animal life the sense of smell appears to play an important role in food finding and it must also be assumed that the same sense is used to bring the sexes together for mating but there has been very little serious investigation of the subject. In many simple aquatic animals there is no proper mating or copulation, the sperms simply being liberated into the water to seek out the eggs, which may be retained in the female's body or may be released into the water. This method is made more efficient if both male and female cells ripen and are shed at the same time. In oysters the shedding of sperms by one male triggers shedding by countless others through some form of chemical stimulation and there are presumably similar co-ordinating mechanisms in other aquatic animals.

With the current trend for investigation of the chemical senses we may expect that the unsatisfactory state of knowledge of its use by the lower forms of animal life will soon be improved. In one particular field, however, there have already been a number of worth-while observations. Starfish prey on a wide variety of molluscs, such as cockles, whelks and limpets. They probably find them by smell and some starfish can dig through sand to find buried clams. The shellfish do not accept their fate passively. Their shells do not provide complete safety from a starfish which will cling to them with its tube-feet and use its muscular arms to pull the shells apart. To avoid this stranglehold, some molluscs extend the fleshy mantle over the shell so that the tube-feet of the starfish cannot get a grip. Other molluscs take evasive action by moving away. Some limpets move away at a speed that may be slow to our eyes but fast enough to escape a starfish. Sand dollars, which are a form of sea urchin, burrow into sand if a starfish comes within 0·5 metres and scallops swim away by clapping their shells together. The spiny cockles leap by flicking movements of their tongue-shaped foot. The movements may be sufficiently violent to throw the starfish on its back and immobilise it for a time.

In most of these cases the starfish has actually to touch its prey to cause escape, so a contact chemical sense must be involved. It has been found that the stimulating substance from the starfish is in its tube-feet, that is, the part that comes into contact with the prey. In whelks, the same reaction to starfish can be elicited by detergents which physically affect the surface of the tissues and it seems likely that the stimulant from the starfish has the same effect on the whelk's tissue. An odd corollary to this came to light after the wreck of the tanker Torrey Canyon off the Cornish coast. The oil that drifted ashore to cover many beaches was dispersed with the detergent BP 1002. Whelks, cockles and queen scallops, which react to the presence of starfishes, were found to be very sensitive to this detergent but certain winkles that ignore starfishes were relatively insensitive. However, after prolonged exposure, even animals of the first group became insensitive to the detergent. Presumably, they were then insensitive to starfish stimulant as well and consequently more vulnerable to starfish attack. After the clearing-up of the Torrey Canyon's oil a number of animals disappeared from the beaches and there is some evidence that this happened because they failed to react to the approach of starfish and so became scarce through predation.

It is usually the case that wherever one animal preys on another, the latter has some means of defending itself. The defence is obviously never perfect but the so-called struggle for survival can be seen as a sort of zoological arms race in which predators develop more efficient ways of tracking and killing and the prey find more effective ways of avoiding being located or captured. Most of our information about the offensive and defensive methods of animals is based on what we can see and hear. Horns, hoofs, fangs, alarm calls and camouflage are well known. Escape reactions elicited by the odours of a predator have already been mentioned but prey animals also use what may be called offensive defence, in two senses. They secrete volatile substances that repel would-be predators from attacking. Often this is not so much an odorous communication as an actual weapon that will harm the attacker.

The best known of all chemical defences is that of the skunks of America. The striking black and white coloration is the skunk's first line of defence. Like the black and yellow stripes of a wasp, the pattern is a warning that the animal has a powerful defence and is best left alone. If the visual warning fails, the skunk raises its tail and discharges a stream of foul liquid from anal glands at the base of its tail, with accuracy over a four-metre range. The spotted skunk gives warning by performing a handstand and also discharges from this position. The liquid smells repulsive; W. H. Hudson described it as appearing to 'pervade the whole system like a pestilent ether, nauseating one until sea-sickness seems almost a pleasant sensation in comparison'. The common chemical sense is also stimulated by the discharge and a skunk attack can cause severe choking and temporary blindness.

## Attracting and repelling

Skunks belong to the family Mustelidae, to which the badgers, weasels and their allies belong. All have anal glands whose original function seems to be for marking territories, as will be discussed in Chapter 8, but several members of the family have followed the skunk in developing the glands for defence and these, too, have a conspicuous warning coloration. The polecat's smell is proverbial, or it was until the polecat became rare, and its old name was foumart, derived from foul marten. The inoffensive pine marten was called the sweet mart. The ratel or honey badger of Africa and the grison of America are other mustelids that produce foul stenches when disturbed. The mongoose family, Viverridae, also uses anal glands for territory marking and some, such as the stripe-necked mongoose, have paralleled the skunks and other mustelids in developing both black and white coloration and strong smell for defence.

Offensive odour is used as a warning by a number of other animals including the grass snake. On being attacked, a grass snake may employ several lines of defence. It can blow up its body and hiss to appear fierce, it may feign death by rolling over and letting its tongue loll out or it may exude a foul stench from glands near the anus.

If these offensive odours are only marginally a form of communication, the same cannot be said for the escape reactions of a number of fishes which are elicited when one of their number is wounded. Rather than the predator being dissuaded from attacking, its presence is communicated to other members of the prey species. The reaction was discovered accidentally by Otto von Frisch while he was working on other behavioural problems in minnows. Von Frisch had performed a simple operation on a minnow and returned it to its tank. Instead of accepting their former companion into the school, the other minnows snapped at it, then fled. For some time they were very timid but eventually they returned to normal. On investigation, it was found that, when the skin of a minnow is broken, special cells are ruptured and an 'alarm substance' is released into the water. This is a pheromone which causes the other members of a minnow school first to flee, then to bunch together. The minnows appear very sensitive and continue to flee at the slightest disturbance until the effects of the alarm substance wear off. The reaction is also triggered visually by the sight of other fishes darting about.

A minnow is not saved by the alarm reaction when it is attacked but its death saves the rest of the school. It is a piece of altruistic behaviour. At one time it was thought that there was a non-altruistic function as well because it had been observed that, although young minnows contain alarm substance, they do not react to its presence in water until they are

*An encounter between dog and grass snake.*
*(a) The dog goes to bite the snake but its jowls wrinkle in disgust as the snake emits a foul odour.*

*(b) The dog is repelled by the snake and jerks his head back. (The snake also feigns death with lolling head and gaping mouth.)*

*(c) The smell makes the dog salivate profusely (Photo by Jane Burton)*

fifty days old. The suggestion was that this would prevent cannibalism; as soon as an adult takes a young minnow in its mouth it would drop it and retreat. However, this does not explain why young minnows should be unreactive and experiments have shown that, in fact, cannibalism is not prevented. The young minnows are swallowed whole, without the skin being broken. Furthermore, it has been pointed out that young minnows form a reservoir of food for use by the adults in hard times so that cannibalism is an advantage to the species at the expense of individuals. A more likely reason for the lack of reaction is that young minnows do not live in schools and the reaction is unnecessary.

Alarm substances are found in a single order of fishes, the *Ostariophysi*, which comprises two-thirds of all freshwater fishes, but the pattern of response varies somewhat according to the habits of different species. Bottom-living tenches swim about excitedly with their bodies inclined at an angle to stir up mud as a smokescreen while cryptically coloured stone loaches and gudgeon remain motionless to escape detection.

Outside the fishes, alarm substances are employed by toads, tadpoles, earthworms and others. Earthworms secrete a mucus when disturbed that causes other worms to retreat. If the mucus is deposited in a burrow, other worms will avoid it for some time. Some ants have an alarm pheromone that is discharged into the air. The pheromones are secreted from glands at the tip of the tail or at the base of the mandibles and, when squirted into the air, they evaporate and diffuse rapidly. When a wood ants' nest is disturbed, the workers squirt formic acid from the abdomen. The acid acts both as a deterrent to the intruder and as a pheromone to alert other ants. All ants within a radius of a few centimetres become aggressive and run about with jaws open. They, too, secrete the alarm substance and yet more ants are recruited to the defence of the nest. The alarm substance disperses rapidly so, when the source of disturbance is removed and the defenders stop spraying, the nest soon settles down. The harvester ants of Florida have a more refined alarm system. They secrete an alarm substance from their mandibular glands. Other ants are attracted by low concentrations which lead them to the centre of disturbance, where the higher concentration elicits a vigorous defence of the nest. If the attack continues, secretion of alarm substances continues and, when the concentration in the surrounding air becomes very high, the ants start to dig into the ground to find shelter. For both ants, there is a neat system for matching the scale of the alert with the size of the disturbance. Only as many ants as are needed for the job are mobilised to defend the nest and, as the emergency passes, they are progressively released to return to their former occupations.

# 4 Homing and hunting

There was once a time when salmon could be found in most of the river systems flowing into the North Atlantic but pollution and damming have destroyed the stocks in so many rivers that efforts are now being made to preserve the few remaining salmon rivers. Salmon feed in the sea and swim up rivers to breed in fresh water, a pattern of behaviour called anadromous (from the Greek for running upwards). Young salmon go down to the sea when they have become smolts, at one to eight years of age. They feed at sea for a number of years and return, as grilse, to spawn. After spawning they set off to the sea to feed again.

For many years it had been believed that grilse returned to spawn in the pools where they were hatched and the belief has been confirmed by tagging smolts on their way downstream. As early as the seventeenth century Isaak Walton recorded in his book, *The Compleat Angler*, an experiment in which young salmon had ribbons tied round their tails on the downstream journey and were later recovered on their return. Then in 1939, nearly 470,000 young Pacific salmon were tagged in a stream flowing into the Frazer River of Canada. The several species of Pacific salmon return to spawn only once before dying but nearly 11,000 of the tagged salmon were recovered on their single upstream journey. Of greater significance, no tagged salmon were found in the neighbouring streams. All that survived the period at sea returned to their birthplace, to the identical stream in which they were hatched even though this sometimes meant a journey of hundreds of kilometres upstream.

How they found their way home remained to be discovered. One ingenious suggestion was that the smolts swim downstream backwards and memorize all the landmarks, rocks and so on, for the return journey. It is now accepted that the sense of smell is used. Professor Hasler, working in Canada, caught 300 salmon as they came upstream. He transported them back to the nearest fork in the river and let them make the ascent again, having blocked the nostrils of half with cottonwool. The salmon with clear nostrils went back to the stream where they had been caught, the rest got lost.

The nostrils of salmon and other bony fishes consist of a pair of U-shaped tubes in the head. Water is passed over receptor cells in the tubes either by pressure from the fishes' passage through the water or by the action of cilia. Being unconnected with breathing, they can be investigated without harming the fish or impairing its activity. It is also relatively easy to irrigate the nostrils with samples of water. Records of the nervous activity in the olfactory lobes of a salmon's brain show that water from the spawning site causes considerable nervous response. Water from a point downstream of the spawning stream causes a slight response but water from other rivers elicits no reaction.

Professor Hasler suggested that the salmon must react to some specific odour in the water; perhaps from waterplants or from the bottom of the pool. It is almost likely to be a mixture of odours from a variety of sources that would give a unique label to a particular spawning pool. Very recently, in 1973, further tagging results suggested that pheromones may, in fact, be involved. It often happens that a few tagged grilse are caught in the 'wrong' river but in one tagging series only six out of 2,300 tagged as smolts in the river Coquet in Northumberland were found in rivers that were not salmon rivers. Many more were found in salmon rivers that were not their own. So it seems that salmon returning from the ocean to breed are attracted to water that contains an odour of salmon. It will be generated by young salmon that have not reached the smolt stage and will belong to later generations of the salmon population coming up river. We must assume that each population at a particular spawning ground has its own odour which stays the same, and can be recognised, from generation to generation but that salmon sometimes get led astray by the odours of other spawning grounds.

There is also some evidence for the pheromone theory from findings at the river Parrett in the Bristol Channel. Breeding salmon were introduced to this river and, during the succeeding years, a large number of salmon were caught in its estuary. Most had come from the neighbouring rivers Severn, Usk and Wye and they had presumably been drawn there by the odour of the newly established salmon population. As yet nothing is known about the nature of the odours that the salmon are picking up at the mouth of their home rivers. The problem of homing by smell is interesting because of the necessity for the animal to recognise one specific odour from among the many in the water. Homing also has to work over a considerable distance, so that whereas we are not surprised that a dog should recognise his neighbours by their scent marks it is remarkable that salmon should detect water from their spawning pool as it flows through the estuary of the river.

There was once a demonstration of water-tasting on the television programme *Panorama* in the days of the late Richard Dimbleby. He introduced a water-taster who worked for a Water Board and invited him to sample water from seven tumblers and suggest where they came from. The taster took a sip, savoured the water in the manner of a connoisseur tasting a more sophisticated beverage and said 'River Thames below Teddington Weir' or whatever it was. He was right every time. It is worth remembering here that the flavour of a fluid, be it wine or water, is really an odour conveyed to the nose but there is no doubt that the taster's recognition of water taste or odour was far superior to that of the ordinary man. We have to imagine how much greater is a salmon's perception of odours that allow it to smell out its birthplace several years after leaving it.

The nostrils of fishes do not contain a vast amount of sensory tissue but this is not necessary for sensitivity. The number of receptor cells in the tissue governs the number of odours which can be distinguished, whereas the ability to sense low concentrations of odour depends mainly on the sensitivity of individual receptors. Eels are known to be sensitive enough to detect a few molecules of pure chemical entering each nostril and it is suggested that only a few drops of pheromone per day need to be carried down an estuary to attract the salmon.

On a smaller scale, the nature of the homing of limpets has also proved difficult to solve. The common limpet has a fixed home which can be seen as a scar on the rock. It usually leaves the home site when exposed by the receding tide and during the night. It spends several hours wandering around the rocks, browsing on algae, and then returns home, settling its shell into the scar to make a perfect fit. How limpets get home is a problem that has not been satisfactorily solved. They do not merely return along their outward route. If lifted off the rock and moved several centimetres, they still get back home and obstacles placed in their way are successfully by-passed. It has been proposed that the limpets are making use of trails left from many previous excursions, so that, in effect, a limpet has many routes radiating from its home. If displaced, it only has to search around for one of these trails to be assured of a safe return. If this is so, the trails must be extremely persistent and they must be polarised, to show in which direction home lies.

When it reaches home, the limpet settles facing a particular direction on the scar so that its shell fits the rock perfectly to make a watertight joint and prevent drying up. This is the reason for homing, as limpets living on flat smooth rocks where a watertight joint can be made any-where, show no signs of returning to a fixed home. It is clearly essential that the limpet positions itself accurately to make the joint perfect. It settles in the correct orientation very rapidly and does not 'shuffle' to fit the shell into the right position, like someone settling into a chair. It is

*Limpets and acorn barnacles cluster around a tiny rockpool. Both are guided to their settling places by forms of the sense of smell. The barnacles settle permanently where barnacles of previous generations have left traces of the cement that anchored them in place. The limpets regularly wander across the rock in search of food but return to the same place when the tide drops (Photo by Jane Burton)*

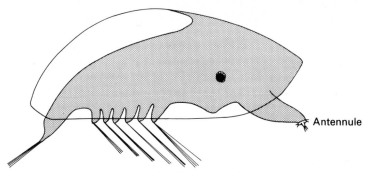

*The body of a barnacle larva is enclosed by the two halves of its transparent shell. When the time comes to turn into the adult form it searches with the paired antennules for minute traces left by other barnacles.*

guided by some kind of 'label' on the scar and a limpet can read the label on scars belonging to other limpets. It will position itself correctly even if its shell clearly does not conform to the shape of the scar. Nothing is known about the nature of the label but it is very likely to be chemical, possibly a secretion from the limpet that is absorbed into the rock. Such a secretion has been found in barnacles, which, although similar in shape and habit to limpets, are crustaceans and not molluscs.

Barnacles start their lives as free-swimming larvae, similar to those of crabs and other marine crustaceans. After a period spent in the open water, the larva finds a solid surface on which to settle. It walks about on its antennules and, at first, wanders in a random fashion. Gradually, it narrows its search, pivoting on one antennule and feeling with the other. After about half an hour, it finally comes to rest. The antennules secrete a cement to anchor the barnacle which never moves again. It now changes into the adult form, surrounded by a box-like chalky shell, and, in effect, standing on its head and waving its legs to collect food.

There is a pronounced tendency for barnacles to settle near other barnacles. This is of economic importance because the fouling of ships' hulls by a coating of barnacles greatly reduces their efficiency. Cruising speed can be reduced by two knots and fuel consumption raised by one half. Researches into fouling have led to an understanding of what makes a barnacle larva settle in a particular spot. The larva explores likely surfaces, moving on if not satisfied and eventually settling near other barnacles or where barnacles have previously lived. It is searching specifically for a layer of protein, no more than a few molecules thick. The protein is secreted by settled barnacles and remains to act as a pheromone after they have died and dropped off. Its action can be demonstrated by rubbing an aqueous extract of barnacle tissue onto a flat surface which then becomes attractive to barnacle larvae. The phero-mone comes from the hard outer shell of the barnacle and may be arthropodin, an important constituent of the outer integument of all arthropods, the huge group of animals which includes insects, crustaceans and spiders. However, arthropodin is soluble in water and the substance that promotes settling by barnacles is insoluble. This is important because as with the label of limpet home sites, there is no question of the barnacle larva being attracted from a distance by an odour emanating from other barnacles and dispersing through the water. The larva only recognises the pheromone when in contact with it. It is intriguing to speculate that the barnacle is 'feeling' the shape of the protein molecules and one wonders whether it is matching the molecular vibrations in its receptors with those of the pheromone or whether the pheromone molecules have a particular shape that fits like a key into sites on the receptor, in accordance with either Wright's or Amoore's theory of the olfactory process.

The use of scent in searching for food has been the least studied. It is easy to assume that an animal is following a scent trail, as for instance when we see a dog running with its nose to the ground, but we have very little idea of what information it is picking up. A few experiments have now been performed to show that some animals do use olfactory clues for food-finding. It might be thought that this is merely proving the obvious. However, the obvious has to be proved before it can be used as the basis for hypothesis. Squirrels, which are famous as food-hoarders, bury seeds and nuts singly over a wide area rather than in a hoard, as is generally believed. They seem not to remember where each item is buried but apparently locate them by smell, even through a layer of snow. There is, however, no proof as yet that squirrels employ smell to find their buried nuts but the ability of deermice to find buried seeds has been investigated in the laboratory. Such a study has wider implications because deermice have disrupted reafforestation programmes by removing planted conifer seeds. The situation is made worse because the commercial felling of trees, that makes reseeding necessary, causes the formation of a habitat where deermice and other rodents thrive. They have been known to remove 100 per cent of newly planted tree seeds. In the laboratory, deermice could find conifer seeds buried under five centimetres of peat in pitch darkness. The only possible inference is that they were able to smell the seeds through the layer of peat.

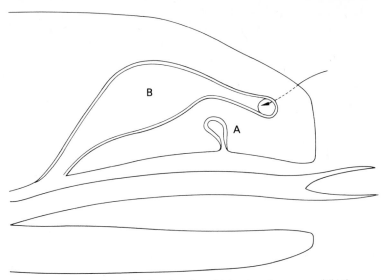

*Section through the head of a lizard to show (a) Jacobson's organ and (b) the nasal cavity.*

It is, again, apparent that smell is used by snakes for finding food. Jacobson's organ is well developed in lizards, as well as in snakes, but it is reduced or even missing in arboreal reptiles such as chameleons and anole lizards. The ducts leading from the roof of the mouth to the sensory tissue of the organ are so placed that the tips of the forked tongue can be brought past them, although there is no conclusive evidence that the tips actually enter the ducts. Reptiles have a well-developed nose as well and there is probably a division of functions between the two organs. The nose may be used for initial detection of faint or distant odours and Jacobson's organ used for more thorough investigation, sweeping up odours from the air or off the ground. Little is known for certain about the use of Jacobson's organ but it is known that rattlesnakes use it to detect king snakes. King snakes eat other snakes; they are immune to their poison and kill by constriction, wrapping their bodies around the victim and suffocating it. When confronted with a king snake, a rattlesnake does not raise its head to strike, which would give the king snake a chance to coil around it, but keeps its head and neck on the ground, raises a loop of the body and lashes out at the king snake. This response is abolished if the tongue is amputated, to prevent the use of Jacobson's organ, but the nostrils left open.

## Homing and hunting

To return to food-finding, the responses of newly hatched rat snakes to plugs of cottonwool soaked in extracts of various small animals have been measured by counting the rate at which the tongue flicks in and out of the mouth. It was found that the snakes responded significantly only to 'extract of mouse' and there was no particular response to worms, slugs, fish, insects or frogs. As the name suggests, the main diet of rat snakes is rodents so it is interesting to find that there is a selective sensitivity to the odour of rodents from the time that they hatch, so that Jacobson's organ must be instinctively programmed.

The trail-following ability of snakes has been demonstrated by allowing other snakes or potential food, such as ants, to move around a narrow track in an experimental arena, then removing them and introducing a test snake. American blind snakes, which are not blind but can do no more than distinguish between light and dark, feed mainly on army ants. Colonies of these ants lead a nomadic life, carrying their eggs and larvae with them as they search the forests for food. While on the move, the advance guards lay a pheromone trail which the main body follows. In a test arena, blind snakes are able to locate and follow a pheromone

The head of the African green tree snake. The nose is assisted in the location of prey by the tongue which collects odours from the air and transfers them to Jacobson's organ, an accessory organ of smell in the roof of the mouth (Photo by Jane Burton)

trail after the ants have been removed. They will also follow the scent trails of snakes of their own species and trails of the opposite sex are followed farthest, which is not surprising. Some male snakes can mate successfully when blindfolded but are unable to do so if the nostrils and Jacobson's organ are plugged, or if the female is coated with vaseline to contain her odour.

A problem that arises from the snakes following scent trails in an arena is how do they know which way to go? It is all very well demonstrating that snakes detect odour trails but, in the wild, this faculty has little advantage if there is a 50 per cent chance of the snake following the trail away from its food. For an animal to follow a scent trail in the right direction it must be able to extract information about the orientation of the trail. The little we know about this information comes mostly from the study of ants. Many ants lay scent trails on their forays from the nest in search of food. When an ant has found a plentiful supply of food, it returns to the nest or to one of the main trails to recruit help. At intervals it presses its sting to the ground and lays a short trail of pheromone. The pheromone evaporates to form a narrow 'tunnel' of odour. Other ants find their way into the tunnel and so make their way to the food. As they return they, too, lay a scent trail. So long as the trail exists, ants continue to make their way along it to the food. When the food runs low, ants begin to return empty-handed. They do not scent-mark and, as more ants return without marking, the trail gradually evaporates. Consequently, fewer ants are drawn along the trail and when the food has gone, the trail disappears. Thus, the food is removed in an economical fashion, no more journeys being made than are necessary. However, if an ant blunders into the trail halfway along, it cannot tell which direction to go, despite the apparent orientation of the trail in the form of an odour gradient. When an ant lays a trail, it presses its sting against the ground to leave an arrow-shaped mark but the following ants are incapable of detecting the shape of the mark and learning the direction that the marker took. Flying insects have an easier task in tracking an odour to its source. They merely follow the odour upwind. If the odour is lost, the insect immediately turns to fly crosswind and casts about until it is found again.

Snails are regular trail followers, a fact that is not surprising as a snail leaves a trail of mucus which is secreted to lubricate the movements of its foot. Marine, freshwater and land snails follow their own trails and those laid by members of their own species. A freshwater pond snail, *Physa*, that is often kept in tropical aquaria, follows its own trail when ascending

## Homing and hunting

to the surface to breathe. Experimentally, at least, one *Physa* will react to the trail of another and follow it in right direction. This reaction is probably used by several kinds of snails to form aggregations on feeding grounds and in resting places. It must also be a way of bringing the sexes together and some snails prey on other snails, finding them by following their trails.

# 5 The odorous world of insects

From the previous two chapters it is clear that our knowledge of the role of smell in the lives of many animals is very uncertain. As yet, information on the subject is scant. We may know no more than the fact that a particular animal has a sense of smell or we have only inconclusive, perhaps even contradictory, evidence for its function. As the scientific investigation of the sense of smell proceeds the picture will become clearer. This has already happened for the insects, a group of animals whose senses have been more thoroughly investigated than most others. The sense organs of insects are well formed yet their behaviour is relatively simple and consequently easy to analyse. Moreover, insects are economically important as pests and carriers of disease so their study has been particularly necessary.

Although the sense organs and nervous system of insects are quite complex, their behaviour is restricted because the sheer size of an insect's body limits the size of its brain and makes sophisticated behaviour impossible. What passes for apparently purposeful behaviour in an insect is merely a blind obedience to orders. The insect is born with a set of behaviour patterns already planned and they are executed and co-ordinated on the receipt of simple commands or stimuli from the environment. In the same way as a military operation is organised so that it can swing into action when the codeword is transmitted so an insect's mating, feeding or egg-laying behaviour, which may be quite elaborate,

## The odorous world of insects

can be initiated by a single stimulus. It may be the sight of food, the sound of the cricket's chirruping or the smell of a cabbage leaf, to give but a few examples.

Many invertebrates possess no more than a common chemical sense but the insects, like the vertebrates, have developed separate senses of taste and smell. The distinction is, however, rather more arbitrary than in vertebrates. There is only one kind of chemical receptor cell, perhaps best referred to as a chemoreceptor but, for land insects, with which

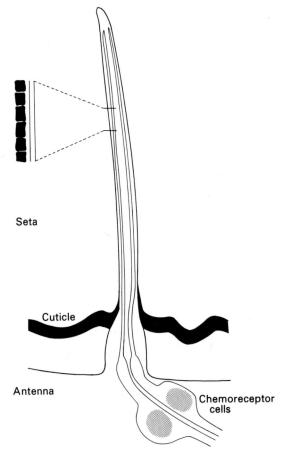

Seta

Cuticle

Antenna

Chemoreceptor cells

*Insect sensillum arising from an antenna. The hollow seta contains the fine processes of two chemoreceptor cells. Odours molecules penetrate pores in the seta, shown in the enlarged section, and stimulate the chemoreceptors.*

we are most concerned, the sense of taste can be restricted to the sensation of dissolved substances while smell is concerned with volatile airborne substances. The basic chemical sense organ in insects is the sensillum, a structure which consists of a hollow hair or seta rising from the surface of the body. It is perforated throughout its length by a large number of pores which connect with a fluid-filled, central lumen that contains the tips of one or more chemoreceptor cells. Odour molecules are thought to be adsorbed onto the surface of the hair and to diffuse to the pores where they pass into the central lumen and so to the chemoreceptors. In some insects, such as bees, wasps and beetles, the hairs are replaced by circular pore plates that lie flush with the surface of the cuticle, like a man-hole cover.

The chemoreceptors are distributed in groups around the body, particularly on the mouthparts and antennae. Those on the mouthparts are particularly concerned with the chemical sensation of food and the antennae are most used for the detection of odours from distant sources. However, the antennae are the site of many senses – touch, air movement, sound, heat and taste as well as smell. It is consequently not possible merely to say that an insect is smelling an object that it is investigating with its antennae. It is receiving a complex array of sensory impressions.

The chemoreceptors are divided into two main types. There are generalists which react to a variety of odours and specialists which are specific to one odour or to a small group of odours. The latter are especially significant in the lives of insects. Blowflies have a 'carrion receptor' which is specifically sensitive to the odours of amines and mercaptans, which are volatile substances given off by rotten meat. As soon as these odours are detected the blowfly moves towards their source, so that a simple stimulus guides the insect to a suitable place for feeding and egg-laying.

Another specialist chemoreceptor has been found on the antennae of a blood-sucking assassin bug, known as the *benchuca* in South America. Charles Darwin suffered from an attack of this bug in 1835 and later suffered a long period of ill health which is now thought to have been Chagas disease, a parasitic infection transmitted by assassin bugs. The chemoreceptors of the *benchuca* are sensitive to human breath but little else. The actual component or components of breath that cause the response are not known but this certainly seems to be the way that the bugs find their hosts, usually as they sleep.

Three generalist chemoreceptors have been found on the antennae of the yellow fever mosquito *Aedes aegypti*. They can be distinguished by the form of the seta and they react to fairly broad classes of substances. One kind is stimulated by fatty acids but its reaction is depressed by essential oils. Another is stimulated by fatty acids with molecules containing over seven carbon atoms but depressed by fatty acids with smaller molecules. The function of the third chemoreceptor is not known, nor is it known what use the mosquito makes of the information about fatty acids and essential oils.

In addition to the sense of smell being widely used by insects for finding food, bringing the sexes together and for locating suitable sites for egg-laying, it is also used for trail-following and nest defence. For the social insects, particularly bees and ants, primer and releaser pheromones are the means for co-ordinating the behaviour of large numbers of individuals and regulating the activities of complex societies. In the control of reproduction and mating, the secretion of pheromones is closely tied to the hormonal control of the reproductive cycle. In one

*The cockchafer uses its antennae to find food. The fan shape of the terminal segments increases the area of the antennae to make them more sensitive. Some cockchafers can smell out truffles, fungi that grow underground (Photo by Jane Burton)*

cockroach that has been well studied the pheromone that the female secretes to attract the male is controlled by the corpus allatum, a gland that regulates egg development. After mating, when the eggs are ready to be laid, the corpus allatum becomes inactive, the pheromone secretion stops and further mating is inhibited. The corpus allatum of a desert locust secretes a hormone which, in turn, causes the secretion of a pheromone that stimulates the growth of the corpora allata in other desert locusts. The result is that a group of locusts mature together and all are ready to mate at the same time. This is one of the factors that cause locusts to appear in swarms whenever conditions are favourable.

The two factors of simple behaviour and economic importance have led the sex-attractant pheromones of moths to be the subject of intensive research. The chemical nature of several pheromones has been unravelled and some have been synthesised with a view to their use as chemical traps. The first sex pheromone to be isolated was bombycol from the silkmoth *Bombyx mori*. Silkworms are raised commercially in large numbers so there was no particular difficulty in obtaining the 313,000 female moths that a team of German scientists used to extract the pheromone. The first stage of this tedious task was to cut the tip off the abdomen of each moth. These were crushed and mixed with water, filtered and the remaining liquor evaporated. From the residue the pure pheromone was extracted and purified. The yield from nearly one third of a million moths was 4 milligrammes of pheromone, that is, a very small drop of liquid. But its potency was enormous. One thousand million millionth part of this ($10^{-18}$ gramme) dissolved in one cubic centimetre of solvent is sufficient to stimulate a male silkmoth. Once isolated, the pure bombycol was analysed and it was found to be an alcohol with the structure:

$$CH_3 (CH_2)_2 \ CH=CH \ CH=CH_3 \ (CH_2)_9OH$$

At about the same time, American scientists were isolating the pheromone of the gypsy moth. Using the same method as the Germans, the Americans were able to obtain the pure pheromone, called gyptol:

$$CH_3 (CH_2)_5 \ CH_1CH_2CH=CH-(CH_2)_6OH$$
$$OCO \cdot CH_3$$

The identification of gyptol was important because gypsy moths had been imported into North America from Europe and their caterpillars were causing severe damage to trees in orchards and woods. In 1869, Leopold Trouvelet, a French artist, conceived the idea of crossing gypsy moths with silkmoths to breed a silk-manufacturing species that would thrive in the climate of New England. The experiment was a failure but the gypsy moths escaped and are now threatening to spread beyond New England. By good fortune it has been found not only that a substitute for gyptol could be synthesised, using castor oil as a raw material, but that the artificial pheromone, gyplure, is more potent than the natural gyptol.

It was hoped that gyplure could be used to stop the depredations of gypsy moths, as the use of pheromones offered a way of controlling insect pests, without the indiscriminate use of chemical poisons which are known to have a terrible effect on harmless animals. In theory, this can be done in three ways. The artificial pheromone can be used to bait traps to catch male moths, which are then killed or sterilised. This method has the great advantage that the poison or sterilant can be aimed exclusively at the pest species rather than broadcast so killing off beneficial or neutral species. Unfortunately, it is difficult to work in practice as the pheromone evaporates quickly in hot and windy weather. Low concentrations fail to attract the moths and high concentrations cause the moths to make mating movements on the surface of the trap, then fly away without entering it. However, the method has been used successfully against the oriental fruit fly on Rota, a small island in the Mariana group. Huge numbers of small boards impregnated with an attractant (methyleugenol) and an insecticide were dropped from aeroplanes and the fruit flies were wiped out.

The second method of control with pheromones is to spray large quantities of pheromone over the infected area, so saturating the insects' olfactory receptors and masking the pheromone given off by the waiting females. A large-scale trial using gyplure was unsuccessful but a small-scale trial on cabbage looper moths has shown promise. About 0·1 gramme per acre of pheromone was needed to prevent mating. The third method has a different aim. The pheromone is used in traps, not to eliminate the insects but to capture samples, estimate the total population of the pest and forecast plagues. When the population is seen to be increasing, the time has come to start control measures, such as spraying with insecticides. The number of males caught is related to the size of the succeeding generation of moths so there is plenty of time to organise a control programme. This method has been useful in helping to check the spread of the gypsy moth because the traps will collect moths as soon as they spread into a new area.

The female gypsy moth does not fly so she needs a signal of some kind to attract males to her. A pheromone is a very practical signal. The sensitivity of the males is such that very little pheromone has to be produced. Secretion of the pheromone starts when the female is ready to mate and stops as soon as she has mated, so that surplus males are not attracted unnecessarily. The pheromone spreads downwind, covering a large area and attracting a mate even if the males are few and far between. The males are stimulated to react when the concentration of pheromone in the air is very small. Their reaction does not change much until the concentration is raised about a million times, so the males are not sent on detours by small eddy currents in the air. They fly steadily upwind until they reach a steep gradient in pheromone concentration near the female. The gradient may only start a few centimetres from the female but the males can steer accurately towards her by ensuring that the two antennae are stimulated equally. It is often claimed that the male moths can home to the females from distances of over one kilometre, but this is based on experiments where marked males have been released at various distances from a captive female and their arrival recorded. It has now been realised that the males were searching at random and only detecting the odour of the female when fairly close to her.

The specificity of a pheromone to a single species is not always so complete as is suggested by the definition given in Chapter 1. The pheromone of the female gypsy moth *Porthetria dispar* also attracts males of the nun moth *Porthetria monacha*, and the cabbage looper moth *Trichoplusia ni* shares a pheromone with the alfalfa looper *Autographa californica*. The two species, in each instance, breed at the same time and share many kinds of food plants, yet they do not interbreed despite using the same sex-attractant pheromone. Laboratory experiments have shown that female cabbage loopers produce much more pheromone than do alfalfa loopers and male cabbage loopers only react to high levels of pheromone concentration. Male alfalfa loopers prefer low levels and female alfalfa loopers never secrete enough pheromone to attract male cabbage loopers. On the other hand, pheromone specificity is extremely definite in some cases. One moth, *Bryotopha similis*, has been separated into two species by its pheromone which exists in two isomeric forms, that is the molecules of the pheromone exist in two mirror-image conformations, identical in chemical composition but different in shape. One species is attracted by the *cis* form of the pheromone and the other by the *trans* form, which are secreted by the respective females. Each form of the pheromone is actually repellent to the males of the 'wrong' species.

Male-attracting pheromones are used by a considerable number of moths. Their use by a particular species can be inferred by inspection of the antennae. Females have simple antennae but the males have feathery antennae, in which the surface area for absorption of phero-mone molecules is greatly increased to form a molecule-catching net. Sex attractants are found in a number of other insects, for example the house fly, chafer beetles and bugs, and there is an interesting differ-ence between moths and butterflies. Both feed on nectar but moths generally feed in dim light and are attracted to flowers by scent. Butter-flies feed during the day and seek flowers by their colours. Moths use scents for bringing the sexes together and butterflies are attracted to each other mainly by visual displays, although some females employ scents. By and large it is the males of the butterflies that carry the perfume, which is often perceptible to the human nose. Meadowbrown and grayling butterflies are said to smell of old cigar boxes. In courtship, the male grayling intercepts passing females which, if responsive, settle on the ground. The male alights and faces her to perform a courtship ritual of opening and closing his wings and waving his antennae. Finally, he clasps the female's antennae between his two forewings and afterwards mates with her. The significance of the clasping action lies in the patches of special scales, called androconia, on each forewing. The androconia are scented and act as an aphrodisiac to make the female receptive to mating.

Male butterflies of the family Danaidae bear a pair of small brushes, called hairpencils, on the abdomen. These are raised during courtship and are used to transfer scented powder from pockets on the hindwings. The male queen butterfly of Florida hovers in front of the female in flight and brushes her head and antennae with the hairpencils so that she is coated with the powder. She alights, the male repeats the dusting and alights to mate. The powder is made up of tiny spheres which contain a viscous glue and an aphrodisiac pheromone. The monarch butterfly, a danaid famous for its spectacular migrations, has very small hairpencils which are functionless. This species courts by pursuing the female, win-ning her by visual displays rather than by perfumes.

Both sexes of the owlet moth secrete scents. The females attract the males who proceed to stimulate them in much the same way as in the danaid butterflies. The scents are secreted from pockets on the abdomen. Usually the lips of the pocket are held tight but, for mating, they are opened and brushes that have been charged with scent from glands lining the pocket are everted and the scent blown away. In the angle shades moth, the female sends out her chemical signal at dawn. When the males reach her they evert their brushes for a moment and then mate. The function of the male's scent is probably to inform the female that he is of the right species and to stimulate her into breeding condition.

The sensory link between courtship and feeding behaviour in the butterflies and moths draws attention to the evolution of the insects, in particular the butterflies, moths, bees, flies and others, that feed on nectar. There is a remarkable parallel between the appearance and diversification of many kinds of insects and of the flowering plants. The first flowering plants were pollinated by wind, which is a wasteful process, but then a partnership was formed between the plants and the insects. The insects act as messengers, carrying the pollen between one flower and another, and they are rewarded with nectar. The process would still be wasteful if the insects moved among the flowers at random, but an individual insect will concentrate on flowers of a single species of plant. That they do this can be seen by watching bees or butterflies at work. The bright colours and scents of flowers that act as signals to the insects may be present only when the flower is ready for pollination and the insects learn which flowers are profitable to visit. Among some of the butterflies, scent stimulates searching behaviour and the colour of the flower guides the search.

For honey-bees and bumble-bees the colour and shape of flowers are important guides. Inexperienced bees recognise the patterns of flowers instinctively but have no inborn knowledge of their scent, which has to be learned. However, once it has visited a particular flower a bee will look for similar flowers and confirm their identity by means of their scent. Both sight and smell may then be used to guide the bee to the nectary at the base of the flower. As early as 1793 it had been noticed that many flowers that were visited by bees had petals with spots or lines of contrasting colour converging on the entrance to the nectary. These became known as honey or nectar guides and it has been shown experimentally with model flowers that they act as leading lines to guide the bees to the nectar.

Such is the dependence of many flowering plants on the visits of insects that some make use of insects' nutritional and reproductive needs to ensure that a given insect will transfer pollen between flowers of only one species of plant. There are the orchids that lure male insects by imitating the appearance of the female insect so that, in attempting to mate with the 'dummy', pollen is transferred from the insect. The bee orchids also liberate an odour that mimics the sex attractant of the female

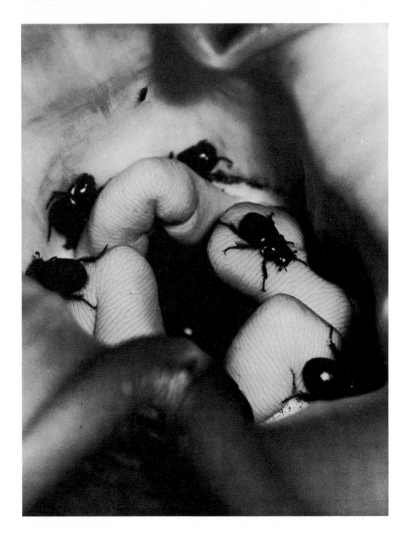

The carnivorous beetle-trap flower of Africa lures small dung beetles by its smell of carrion. The beetles become trapped because they cannot climb the smooth waxy petals. They eventually die and are digested by the flower (Photo by Jane Burton)

bees and the odour of some of the orchids is more effective than that of the female insect herself. Then there are the flowers that produce odours that imitate the odour of the insects' food or egg-laying sites. Blowflies are attracted to lay their eggs in the evil-smelling flowers of the African carrion-flowers, and the cuckoo-pints have a cunning trap to ensure cross-fertilisation. They emit an odour of dung and attract dung-eating flies and beetles, which slide down the slippery sides of the ensheathing leaf. The insects are trapped by a ring of spines and, during their attempts to escape, any pollen they are carrying is brushed onto the female flowers. After a while the male flowers ripen and the insects pick up new pollen. The barrier of spines wilts and the insects escape to fly about until they are lured into another cuckoo-pint.

However, the most ingenious olfactory link between plants and insects is seen in certain orchids which secrete oils rather than nectar. The plants are making capital from the territory-marking behaviour of solitary bees. Male bumble-bees have marking places scattered around their territories where they deposit scent to attract females. The scent is produced in mandibular glands and spread by chewing the edge of a leaf or post. Male *Euglossina* bees of Brazil do not manufacture their own scent but collect it in the form of oils from these orchids and store it in spongy receptacula on the hind legs. They are attracted to the orchids by the smell of the oils and pollination is carried out as the bees move from orchid to orchid collecting oil. When the bee's receptacula have been charged, it releases the scent into the air by fanning its wings to increase evaporation and females are attracted to it.

The use of odour for food-finding has been touched on in the examples of plants mimicking the food odour for the purposes of pollination. The odour that naturally draws an insect to its food is merely a by-product of the food-object's metabolism. Houseflies and their relatives are drawn to flowers and dung by scent. Fruit flies, which under their generic name of *Drosophila* are better known as the subjects of genetic research, feed on the juices of overripe fruit. Some species have become pests and the Mediterranean fruit fly threatened the existence of orchards when it was imported into Florida. Fruit flies are guided to rotting or overripe fruit by the alcohols and acids produced by fermentation and the attraction of these substances explains why fruit flies become a nuisance by gathering around alcoholic beverages and pickles, the vinegar used for pickling being a weak solution of acetic acid.

Chafers are heavy-bodied beetles with fan-shaped antennae. Most feed on leaves or petals or suck sap and nectar and can be pests when abundant. One species hunts truffles, underground fungi beloved by gourmets, which are found with the aid of dogs or pigs trained to smell them out. The chafer can detect the position of a truffle while in flight. It will suddenly stop, drop to the ground and burrow down to the truffle.

There is a broad overlap in the requirements of food and egg-laying among insects. For many species the larval stages have the same food requirements as the adults so the eggs are laid where the parents feed, but this is not so for the butterflies, where the caterpillars eat leaves and the adults sip nectar. In this instance, two sensory systems are needed, one to locate the adult's food supply and the other to ensure that eggs are laid where the larvae can feed. The white butterflies react to red, yellow and blue surfaces, the colours of flowers, with feeding movements but make egg-laying movements only on green, the colour of leaves. Presumably there are insects with similar behavioural responses mediated by olfactory clues.

The sensory appreciation of suitable substrates for egg-laying has been studied in several insect parasites as they form a field of economic as well as intrinsic interest. The ichneumon flies and chalcid wasps are solitary-living relatives of the bees and ants in the order Hymenoptera. Their eggs are laid in the adults, larvae and eggs of other insects and the larvae feed on the tissues of the host. The chalcid wasps even lay their eggs in the larvae of ichneumon flies that are, themselves, living inside the body of a caterpillar. Two ichneumon flies, *Alysia* and *Mormoniella*, find their hosts in carrion. *Alysia* is attracted by the smell of the meat and lays its eggs in the maggots of blowflies feeding there. *Mormoniella* has slightly different preferences. It lays its eggs in the pupae of blowflies. Its sense of smell is programmed to arrange that it goes to meat that is several days old, old enough for the blowfly larvae to have pupated. Fresh meat gets little or no response from *Mormoniella* and the full response is only elicited when the meat is over nine days old.

Another ichneumon fly, *Opius*, parasitises the larvae of the melon fly *Dacus curcubitae* that lives in fruits and is a pest in Hawaii. Females of *Opius* are guided by scent to fruits but then need the scent of the larvae to stimulate egg-laying. For the first three days after emerging from the pupa, female melon flies are not attracted by fruit. Their sensitivity only develops when their ovaries ripen.

Tracking down a suitable host requires more than a single sense and the selection of a host has been thoroughly investigated in *Trichogramma*, a minute chalcid wasp that lays its own eggs in the eggs of 150 different kinds of insects, including moths and weevils. The eggs are detected from some distance, by odour in some species, and the chalcid investigates them with its antennae. Two senses are involved in this act: touch and smell. To be suitable, the host egg has to be of a certain size, upstanding and separate from other eggs. It also has to be free of the smell of other *Trichogramma* females. This is important because there is not room for two larvae to develop in one egg. When *Trichogramma* investigates an egg, it climbs over it, so that traces of odour remain on its surface. The next *Trichogramma* to investigate it picks up the odour and immediately rejects the egg. It may sometimes make a mistake but there is a third, unknown but perhaps chemical, sense involved. The female may start to drill into the egg with its ovipositor but it is quickly withdrawn and no egg is laid if the host egg is already parasitised.

# 6 The social insects

Most animals lead solitary lives. An individual spends its life on its own. It associates with members of its own species only for mating and this contact is usually of short duration. The partners separate and, after giving birth or laying eggs, the female has little or nothing to do with her offspring. Exceptions to this immediately spring to mind and there are two groups of animals in which a social way of life has become highly developed. These are the mammals and the insects. The simplest forms of social life can be seen in a wide variety of insects, starting at aggregations for feeding where the animals may still be behaving as individuals and have gathered solely to make use of a localised resource. In a more permanent aggregation there are bound to be interactions between individuals and communication between them is needed.

The flour beetle *Tribolium confusum*, so named because it is readily confused with a near relative, lives in large numbers in stored grain or in grain products such as flour, corn meal and dog food. Its larvae are the well-known mealworms. Flour beetles are a pest, but they have also proved very useful as subjects for the study of growth of populations. A box of flour provides the beetles with all their needs and the beetles can be counted simply by sieving them out of the flour. So it is possible to express the population of beetles at any moment as numbers per unit weight or volume of flour. When flour beetles are introduced to a box of flour their numbers increase rapidly until the population becomes

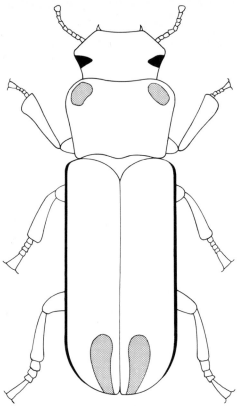

*A flour beetle showing the quinone-secreting glands on the thorax and abdomen.*

very dense. Then the birthrate declines and the population levels off. Part of the decreased birthrate is due to the beetles eating their own eggs as they roam through the flour but this probably has little overall effect. More important are the interactions between a crowd of adult beetles. When flour beetles are disturbed they discharge a mixture of quinones from glands on the thorax and abdomen. The continual jostling of beetles in a dense population is sufficient disturbance to cause the secretion of quinones which kill or deform the larvae and pupae. The natural excretions of the beetles also play a part in limiting the population as the reproductive rate declines when flour from an 'old' colony is mixed into a 'new' colony. The excretions constitute a primer pheromone that acts on the beetles' reproductive system and reduces fertility.

Within these aggregations, flour beetles lead isolated lives. They are living communally but there is no structure of social organisation. True social life is exhibited where a family keeps together to form a communal nest and to share breeding and feeding activities. The beginnings of a social life are seen in several kinds of insects. The female earwig, for instance, lays her eggs in a cavity under a stone and broods them until hatching. She stays with the young earwigs for some time but eventually the family disperses. The significance is that two generations of earwigs have met in a communal 'home' but in only two orders of insects has this situation developed with the family staying together to produce a society in which there is a division of labour with individuals dependent on one another throughout their lives. The orders Isoptera, the termites, and the Hymenoptera, which includes the bees, wasps and ants, have developed social life to such a degree that a colony of one of these insects is sometimes regarded as a 'superorganism' in which the individual members are wholly subjugated to the life of the colony. They cannot live by themselves, any more than a tissue cell can live outside the body of an organism and like the cells of an organism, the members of an insect colony are specialised for particular tasks, such as egg-laying, foraging, guarding and nest-maintaining. The analogy can be overworked but the colony, like an organism, has to control its constituent parts so that their activity is integrated and the life of the colony is ordered. In the insect colony much of the control is effected by pheromones, chemical signals that play a part analogous to the hormones that control the internal activities of an organism. Primer pheromones control the overall life of the colony and releaser pheromones control the day-to-day maintenance activities such as feeding and defence.

There is a gradation of social life in the bees, from the solitary bees in which there is no social life apart from mating to the hive- or honey-bee whose colonies persist from year to year. The rudiments of social life are found in small bees of the genus *Halictus*. The female digs a burrow and lays her eggs in brood-cells that are provisioned with pollen. Some species stay with the brood to guard it and the female may meet her offspring when they emerge. *Halictus malachurus* has progressed to the next stage of social life. All the offspring from the first brood of the year are female but they are infertile like honey-bee workers. They stay with their mother to construct more brood-cells and to collect food while she concentrates on egg-laying. Towards the end of the summer she lays eggs which develop into fertile females and males. Only the new generation of fertile females survive the winter to continue the species.

In *Halictus malachurus* there is a division of labour in the nest and the formation of two castes: the reproductive queen and her infertile workers. The colonies of the honey-bee are an elaboration of this system. A colony may contain up to 60,000 workers which, to keep the nest thriving, have to share amongst themselves the duties of food-collecting, making and provisioning brood-cells, caring for the larvae, regulation of temperature in the nest and guard duties. The queen's only activity is to lay about 1,500 eggs per day. For this packed community to run efficiently there needs to be a degree of control over the activities of each individual so that its behaviour is directed to serving the community's best needs. Control is effected by visual, tactile and olfactory stimuli. Of the three, olfactory stimuli may prove to be the most important, although the complete repertoire of chemical, or other, signals employed by honey-bees has yet to be discovered.

The central figure of a beehive is the queen, the fertilised female who devotes her life to egg-laying. She is more than an egg-laying machine, however, because her physical presence is essential to the wellbeing and continued functioning of the hive. If the queen is removed, the workers become aimless and life in the hive runs down. She acts as a pacemaker for the community and her control over the hive is maintained, at least in part, by a pheromone known as 'queen substance', which is composed of two very similar substances: 9-oxodecenoic acid and 9-hydroxydecenoic acid. Queen substance is produced in the queen's mandibular glands and she smears it over her body. It is licked up by the workers that attend her and passed round the colony by the process of communal feeding. Food brought into the nest by foragers is passed from one insect to another. One regurgitates food from its crop, another licks it up and later regurgitates some of it to a third and so on. Queen substance is transferred at the same time and its spread is enhanced because it is very attractive to the workers and is sought keenly.

In the hive, queen substance acts as a primer pheromone. It prevents the ovaries of workers, which are stunted females, from developing and also prevents the workers from building the specialised brood-cells in which new queens develop. At certain times of the year new queens do develop, apparently because the old queen is becoming worn out or because overcrowding dilutes the supply of queen substance. The process can be stimulated artificially by removing the old queen. The workers, lacking the inhibition of the queen substance, start to rear new queens by feeding a rich diet to larvae that would otherwise have become workers. These larvae produce a pheromone that keeps the colony functioning in the absence of a queen.

When the young queen emerges she either takes over the egg-laying duties of the old queen or leads a swarm of workers to found a new colony. In either case she has to be fertilised. She leaves the hive on a warm afternoon and sets out on a nuptial flight where she is sought by the drones who patrol special areas, flying 10–15 metres above the ground. The queen's presence is relayed on the air to the drones by queen substance, which now acts as a releaser pheromone, stimulating their sexual behaviour and guiding them to the queen. The drones can pick up the scent of queen substance from a distance of about 50 metres and when within 10 metres they can pick her out visually. Each queen mates with several drones, and she receives sufficient sperms to fertilise the many thousands of eggs that she will lay in her lifetime.

After her nuptial flight, the young queen gathers a swarm of workers and takes her leave of the old hive. They settle on a near-by branch while scouts search the neighbourhood for a suitable place in which to found the new nest. The importance of the queen and of queen substance can be demonstrated easily at this stage. While in the air, the swarm of workers keeps close to the queen. If she becomes separated they search for her and, if she is removed from the settled swarm, the workers become restless. They run about the surface of the swarm or fly around in the immediate vicinity in search of her. A queen placed inside a gauze cage will keep the workers quiet but a queen kept in a polythene bag, so she can be seen and heard but not smelt by the workers, is treated as if she were not there.

When the bees have settled in their new home they acquire a distinctive nest odour which is compounded of the fragrances of the flowers that they have been visiting for food. The process of food-sharing ensures that the odour is thoroughly mixed and distributed through the colony. The nest odour enables the bees to recognise strangers and when a beekeeper finds it necessary to implant a new queen in a hive he has to ensure that she is not killed by the workers by placing her in a cage whose entrance is plugged with sugar. By the time the workers have eaten the sugar, the queen will have gained their nest odour and can safely enter the hive. So specific is this odour that neighbouring hives have separate odours even though the workers will have been collecting nectar and pollen from the same flowers. The specific nature of the nest odour helps to guide bees back to the hive and it is aided in this by a pheromone that is deposited by the feet of bees as they walk through the entrance of the hive.

If a bee attempts to enter the wrong hive it will be recognised as a stranger by the guard bees standing at the entrance. The intruder may retreat when challenged but if it persists in trying to enter the hive it will be stung. Mice and other hive robbers are treated in the same way. At the same time as the guard bee uses its sting, it liberates iso-amyl acetate from glands in the sting apparatus. This is an alarm pheromone which alerts other guards to the danger and gives them a target at which to direct their stings.

The most important function of worker bees is to collect food for the queen, the larvae and for their nest-bound colleagues. The collection of pollen and nectar from flowers is made efficient by the bees being able to communicate the whereabouts of a suitable source of food to other workers. The floral odour that clings to a bee when it returns to the hive alerts other bees and the bee also signals that it has foraged successfully by exposing the Nasanoff glands on its back to liberate a scent.

The way that honey-bees communicate the position of a food source to other bees was the subject of a classic investigation reported by von Frisch in 1946. Von Frisch trained bees to feed at dishes of sugar water and showed that other bees learned the position of the dishes from the behaviour of the trained bees when they returned to the hive. The 'language of bees' has now become almost general knowledge, so a brief summary only is needed here. Von Frisch demonstrated that the home-coming bee informed the other bees of its find by 'dancing' on the comb. If the food is less than eighty metres away it merely walks in a circle (the round dance). For a distant source, the bee walks in a figure-of-eight (the waggle dance). As it comes down the centre of the figure-of-eight it waggles its abdomen to show the distance to the food: the slower the waggle, the farther the food. The orientation of the dance indicates the direction of the food. If the bee 'waggles' vertically up the side of the comb, the other bees must fly towards the sun to find the food. If it 'waggles' downwards, the food is to be found by flying away from the sun. When the 'waggle' is performed at an angle to the vertical, the foraging bees choose their course by laying off that angle from the bearing of the sun.

The discovery of 'waggle dances' of the bees is one of the most famous stories in the history of zoology. It is to be found in every textbook but doubts are now being expressed and there is a suggestion that bees are communicating the position of food by olfactory rather than by visual signals. It takes a brave man to challenge anything that has become an established belief and as von Frisch's findings have been repeated and confirmed by other experimenters, any challenge must be based on very strong evidence. Two Californian biologists, P. H. Wells and A. M. Wenner, consider that they have the necessary evidence to show that the waggle dances are not a source of directional information.

In their experiment, Wells and Wenner set up three feeding stations, each 200 metres from a hive and standing in a semicircle around it. At stations 1 and 3 ten bees were marked and fed on scented sucrose solution and any further bees that arrived were caught and kept prisoner. On the next day stations 1 and 3 were set up with *unscented* sucrose and scented sucrose was given only at station 2. If the direction to the sucrose had been communicated by the twenty bees marked on the first day, foragers on the second day should have gone to stations 1 or 3, but this did not happen. 1,224 new foragers visited the scented sucrose at station 2, but only twenty-five went to station 1 and eight to station 3. It seems that the other bees had not been told where to look for food but had only been given the information that there was a supply of scented food somewhere in the neighbourhood. They had picked up the scent from the marked bees and set out in search for it.

Wells and Wenner point out that experimental evidence cannot prove or disprove a theory; it can only lend it support. They suggest that the evidence cannot support the 'waggle dance' theory. The drawback to investigating the foraging behaviour of bees is that it is difficult, if not impossible, to follow an individual bee. Von Frisch had shown that after one bee had fed at a dish it was followed later by a number of other bees but he did not, and could not, record how many bees had searched vainly in other directions, nor how long it took them to find the test dish. In fact, the evidence is that it takes the bees a far longer time to find the dish than one would expect if they had been given the information to guide them straight there. So they must be searching blindly with only the knowledge of the food's odour to guide them.

It has long been known, however, that, when the source of food is within eighty metres of the hive, the bees find it by smell. The round dance alerts the bees and they obtain the scent clue from the dancer. Perhaps the waggle dance does no more than tell them to search farther away. Natural sources of food consist of sizeable beds of flowers which are relatively large olfactory targets and the bees are not expected to home accurately to a single bloom. But when a bee has found food it will expose its Nasanoff gland at the spot and so manufacture its own olfactory target. As drones can find a queen from sixty metres by scent the queen must manufacture an olfactory target of sixty metres radius. If a single worker bee emitting its scent at a dish of sugar forms a target of the same size it will not be difficult for other workers to find it during a search from a hive 200 metres distant.

Whether honey-bees have a language of dances or rely on recognition of scents to find their food is likely to be debated for a considerable time but some of their less advanced relatives, the stingless bees, are definitely known to use scent marks to guide foragers. Stingless bees lay trails similar in effect to those of ants but, as bees fly from food to nest, they cannot leave a continuous trail in the same way as the earthbound ants. Instead, they deposit a series of discrete marks. When a stingless bee has found food and has visited it several times she returns to the nest, pausing every metre or so to rub her mandibles on a blade of grass or a stone. In some species it is possible for the human nose to detect the characteristic odour of the mandibular gland secretion on the marked object. The scent trail does not go right up to the hive but the returning bee alerts the occupants of the hive by running excitedly in zigzags and buzzing loudly. Samples of the food are also given to other bees and, eventually, a group of bees will leave the hive with the one that brought the news of food and follow the line of marks back to the source.

As research into chemical communication among social insects progresses it may transpire that the termites and ants have a more elaborate complex of pheromones than the honey-bees. That the workers of some species are blind, or can do no more than distinguish light from dark, makes it more likely that the chemical senses are well developed. Among termites, the development of workers is suppressed by the reproductive termites secreting a pheromone like the queen substance of honey-bees. Termites also have a trail-marking odour, but it seems that ants have at least ten pheromone systems. Used singly or in conjunction they order the life of the ant colony. Trail-marking pheromones and alarm pheromones secreted by the mandibular glands or from glands near the sting (Dufour's glands) have already been described in a previous chapter. In the American fire ant the trail-marking pheromone has a function additional to that of leading workers to a source of food. High concentrations cause a large part of the colony, including the queen, to leave the nest and migrate to form a new colony and there is evidence that the queen ant, like the queen bee, regulates the development of the larvae and the behaviour of the workers. It is not yet certain whether her orders are transmitted by pheromones in all instances but extracts from the queens of red ants have been found to affect the growth of the larvae.

One specific ant pheromone can be demonstrated quite easily and

*A colony of driver ants crosses a forest road in Nigeria. The leading ants lay a scent trail that the main body follows blindly (Photo by Jane Burton)*

has been the subject of an amusing experiment. When an ant dies it is treated at first as if it were still alive but, after a day or so, the corpse is picked up and dumped in a refuse heap outside the nest, in response to certain substances formed by decomposition of the body. If a live ant is smeared with these substances, it is immediately picked up and thrown on the refuse heap by its fellows. It makes its way back to the nest and is immediately thrown out again. The performance continues until the 'death pheromone' has worn off its body.

Another pheromone regulating the life of an ant colony is secreted by the larvae. In wood ants, at least, this pheromone stimulates the workers to care for the larvae, grooming them and carrying them to safety when they become displaced. It has been mimicked by a certain beetle which spends its life in ant nests as a parasite of the community. Animals intruding into an ant nest are usually attacked and driven out or killed but there is a variety of animals that make a living in ant nests by sponging food. They include beetles, flies, mites, woodlice, wasps and butterflies and they survive among the ants by managing to be ignored, by defensive adaptations, such as a thick skin, or by mimicry. Most ant mimics have evolved body forms that make them look remarkably like the host ants but the beetle *Atemeles pubicollis*, which lives in the ants' brood chamber, mimics the larvae chemically.

The adult beetles lay their eggs in the nests of the wood ant *Formica polyctena*. When the beetle larva emerges it secretes a pheromone from paired glands on each segment of the body and worker ants are attracted to groom the larva intensively, working along its body. When they reach the head, the larva rears up and is fed by the ant. By tapping the ant's lip the beetle larva imitates the begging actions of an ant larva and, by soliciting vigorously, the beetles get more food than the ant larvae, upon which they also prey. Eventually the beetle larva pupates and, when it emerges as an adult at the end of the summer, it has to find a new home because the wood ants stop breeding during the winter. The beetle begs a final meal by tapping an ant with its forelegs, the same signal as is used by the adult ants themselves in communal feeding, and migrates to a nest of the insect-eating *Myrmica* ants. These ants live in grassland and tend broods through the winter. The beetle is guided to grassland from the woodland home of the wood ants by increasing light intensity but then finds a *Myrmica* nest by its odour. To gain admittance to the colony it draws the attention of an ant by tapping it with its antennae and raising its abdomen for the ant to lick secretions from glands at the tip. The secretions suppress the ant's aggression and the ant is next attracted to more glands on the side of the beetle's abdomen which apparently emit an odour that mimics that of the host ants. Recognition and acceptance is now complete and the ant picks up the beetle and carries it to the brood chamber.

# 7 Smell in birds and mammals

The two most advanced classes of animals, the mammals and the birds, share a common descent from the reptiles. The birds have taken to an aerial life, where good vision is of the utmost importance for the control of flight but where smell is of little use and is rarely used even for locating and recognising food. Mammals, on the other hand, retained the reptilian habit of creeping over the ground. The sense of smell came to play an increasingly important role in food-getting and in social life as they literally nosed about in the undergrowth, often at night, where eyes are of limited use.

Such birds as have been found to make use of their sense of smell use it solely for food finding and not for social communication; there is no evidence for the possession of pheromones by birds. Dissection has shown that several unrelated species have a well-developed olfactory epithelium or olfactory lobes of the brain. These include the turkey vulture, a relative of the condors, the albatrosses, the snow petrel of the Antarctic, the kiwi, the honeyguides and the grebes. In very few instances is there any proof that these birds use their sense of smell, but this may be because they have not been sufficiently studied. There are anecdotal stories of petrels gathering at hot animal fat poured on the sea and these birds also gather around whaling ships to feed on the carcasses but there is no proof that they are attracted by the smell of fat or blubber. There is stronger evidence for the use of smell by the honeyguide, a relative of the woodpeckers, that raids bees' nests for the wax combs. A sixteenth-century Portuguese missionary reported that, when he burnt beeswax candles, his church was invaded by honeyguides.

Since the time of Darwin there has been conflicting evidence as to whether turkey vultures, and other carrion-eating birds, can find food by the odour of putrefaction. Darwin and other experimenters found that rotting meat was ignored if hidden from sight. Turkey vultures would walk over offensive offal covered with thin canvas, never finding it unless a rent was made in the canvas. However, American gas engineers have found that turkey vultures gather around leaking gas pipes, presumably attracted by the odour of the gas, and, in an experiment, turkey vultures were attracted to the fumes of evil-smelling mercaptan which was released from a hidden position at the bottom of a canyon. They descended into the canyon to investigate the source of the odour.

The kiwi has also been the subject of experimental investigation. In some respects it leads a life typical of many mammals. It is nocturnal and skulks in dense undergrowth, rarely exposing itself to view. Earthworms are the kiwi's main food and it finds them by probing in the soil with its long bill. Many other birds, such as starlings and waders, feed by probing for buried animals, but they are located by touch or by sight – the bird peering down the length of its bill as it opens a hole in the ground. The kiwi is unique in having nostrils at the tip of the bill, where they are best placed for smelling out earthworms, but no one seems to have explained how the kiwi prevents its nostrils from becoming plugged with earth. The kiwis' ability to find earthworms by smell was proved by training them to search for food buried in aluminium tubes. Then, some tubes were baited with earthworms and others left empty. They were covered with fine nylon mesh and it was found that the kiwis thrust their bills through the mesh of only those tubes that contained worms.

By contrast with the birds, mammals have a very well-developed olfactory system as is seen in the size of both olfactory epithelium and olfactory lobes. The importance of smell has only diminished in the aerial bats, the sea-going whales and in the more advanced of the basically tree-dwelling primates. The bats have given up the use of smell for the same reason as the birds but there would be no reason why the marine mammals should not be able to smell waterborne odours, as do fish, if it were not for the fact that they have to keep their nostrils closed when diving. One can imagine that if seals and whales had evolved a method of closing the nasal passages at the level of the throat, instead of at the nostrils, it would have been possible for them to allow water to circulate in the olfactory clefts and so to smell waterborne odours. But this did not happen and the sense of smell in whales has atrophied to the point of extinction, the toothed whales – dolphins and sperm whales – having no trace of olfactory lobes in the brain. Seals retain their sense of smell for use on land during the breeding season. Among some tree-dwelling mammals, the sense of smell is still very important. It is used for communication by squirrels, lemurs, tree-shrews and the marsupial flying phalanger or sugar glider but has become progressively reduced in monkeys, apes and Man. There is a parallel reduction in the use of smell by arboreal reptiles such as the chameleons and anole lizards.

That mammals make extensive use of the sense of smell hardly needs to be stated; the domestic dog is so obviously an animal that lives in a world of smells. Yet it is also clear that Man, another mammal, is largely excluded from the dog's world and it is only through observation and experiment that we can get a secondhand impression of what this world is like. First, the size of the dog's olfactory apparatus gives an indication of the extent of its world of smells. The olfactory epithelium of a medium-sized dog has an area about fifty times that of a man and it may contain over 200 million olfactory receptor cells. From what we know of the mechanism of receptor cells, the greater number will allow the dog to detect a wider range of smells than can a man and will also make the dog's nose more sensitive to weak smells because there will be a greater chance of odoriferous molecules stimulating sufficient receptor cells for a message to be transmitted to the brain. The dog also has the advantage of its nose having a better internal shape for allowing air to circulate around the olfactory membrane and its brain is better equipped to analyse the information it receives.

Sensitivity to smells has been tested by blowing a fine spray of a scent, at a known concentration, directly into the olfactory cleft. The performance of dogs can then be directly compared with that of Man and it has been found that a dog can smell a variety of substances at concentrations one thousand to one million times lower than a man can. A sensitivity in the order of one thousand times greater, such as for ethyl mercaptan, is probably due to the ease with which molecules get to the receptor cells in a dog's nose but for a millionfold difference, as for acetic acid, there is probably a difference in sensory equipment, in both nose and brain, between dog and Man.

A dog's sense of smell can be more readily appreciated by behavioural tests rather than physiological experiments but in designing such experiments there is always the difficulty of ensuring that the animal is responding to the right stimulus. There is the danger of it responding to unconscious signals given by the experimenter. He may, for example, look relieved when the dog is about to behave correctly. The dog perceives this and so knows how to behave. In following a trail, the dog may well be using visual clues from trampled vegetation or footprints but as long ago as 1885, G. J. Romanes designed a test to overcome these problems. He set off at the head of a column of twelve men walking in single file. Each man carefully placed his feet in the footprints of the man in front. After an interval the procession split into two halves, each continuing its own way. Shortly afterwards, Romanes's dog was released to follow the trail. When she got to the junction, she unerringly followed the route taken by the six men headed by Romanes. Seventy years later H. Kalmus carried out similar tests involving identical twins. If the dog was given the scent of one twin it would happily follow a trail laid by the other. So the two must have had very similar scents but there is sufficient difference in their scent for the dog to distinguish the two twins if given both scents at once. From this evidence it can be seen that a dog's perception of smells is on a level with our powers of visual discrimination. We can distinguish identical twins when seen side by side but not confronted with one alone.

The trail followed by dogs consists of minute quantities of sweat. A person's sweat contains a variety of dissolved salts and other substances, notably butyric acid. At each footstep, sufficient traces of sweat are left for the human nose to smell the imprint of a bare foot on a sheet of clean blotting paper. A dog can detect the sweat that has percolated through the sole of a shoe, or even through a rubber boot, but it cannot do so until the shoe has been worn for a couple of days to become impregnated with the wearer's scent. The various odours left in a footprint evaporate or sink into the ground at different rates so that the dog can work out the direction of a trail after following it for no more than twenty metres but, when the trail is too old, there is insufficient scent left to guide it. At best, a dog can follow a trail on a hard surface no more than half an hour after it was laid but a trail on grass will last up to four hours.

The use of scent by wild members of the dog family when hunting is perhaps taken for granted but it is only recently that zoologists have seriously investigated the habits of animals in the wild. Groups of animals have been followed for long periods of time and the details of their feeding habits, their mating and upbringing have been recorded. Such observations are often given greater strength because the animals under surveillance can be recognised as individuals and even given names. It may take weeks of practice before this is accomplished but it ensures that the record of a sequence of behaviour has been performed by a single individual and is not an interpretation of a mosaic of events by several animals. *Innocent Killers* by Hugo and Jane van Lawick-Goodall is the account of such a study. Their hyaenas, jackals and wild dogs lived in the plains of East Africa. Smell, hearing and vision were used to find food and, at least for daylight hunting, it seems likely that sight is the most effective sense for hunting large animals on open plains. In a more technical volume, *The Wolf*, L. David Mech records detailed observations of how wolves found their prey in Isle Royale National Park, Lake Superior. They had three basic methods of finding prey: chance encounter, tracking and scenting. The first method is self-explanatory and tracking involves following the trails of the prey. Only very fresh trails interested the wolves

*A hedgehog shows its wet nose as it uncurls. Its eyesight is weak and the sense of smell is very important when it forages through the undergrowth at night (Photo by Jane Burton)*

and scent was probably involved. Scenting involves detecting the odour of the prey as the wolves passed downwind of it. Like the male silkmoth looking for a mate, the wolves turn upwind and head straight towards the prey. Mech reports that the wolves usually detect moose when they come within 300 metres but one cow moose, with twin calves, was scented from two and a half kilometres.

It seems most probable that smell is of most value to a carnivorous animal when searching for small animals tucked away under leaf litter or buried in the soil. Hedgehogs and badgers must depend on smell, and on chance encounter to some extent, when rooting through the undergrowth or herbage at night. Hedgehogs can detect a beetle at one metre but in a laboratory study of feeding preferences hedgehogs would react only to moving prey animals. The use of smell in detecting danger is more apparent than its use in food finding. Badgers will forage happily within a few metres of a man's feet in daylight but retreat as soon as they catch a whiff of his odour. Hedgehogs can smell a man or dog at about ten metres.

For all the importance of the nose to mammals as a detector of food and enemies, it is the role in social communication, that is, the language of smell, for which the sense of smell has reached an advanced, and probably unique, state of refinement. This use has been overlooked to a great extent because of our failure to appreciate it but the scent signals of mammals are as complex and as varied, if not as spectacular, as the songs or eye-catching displays of birds. Like the specialised plumage of birds, a variety of scent-spreading devices have been evolved by the mammals to provide social signals that indicate the identity and state of mind or intentions of the transmitting animal. The information is transmitted either by the scent being deposited as a mark for other animals to investigate at a later date, a form of visiting card, or through direct sniffing of the body by another animal.

The scents, which are pheromones, are based on body secretions – urine, faeces, saliva and the products of skin glands – which carry the specific odour of the individual. When skin secretions are used as pheromones they are produced from enlarged skin glands. Mammals possess two kinds of skin glands, both of which have been important in the evolution of the order. Sweat glands are needed to produce a flow of watery fluid whose evaporation protects the body from overheating. Sebaceous glands secrete a fatty substance around the base of each hair which helps maintain its condition. From these glands, there have evolved the milk-producing mammary glands, which are unique to

*Maxwell's duiker shows its facial glands. Secretions from these glands are used to mark leaves and twigs and act as signals to other duikers (Photo by Jane Burton)*

mammals, and a variety of specialised scent-producing glands. To name a few, there are the chin glands of rabbits, facial or preorbital glands of antelopes, anal glands of beavers, circumgenital glands of marmosets and the tail gland of foxes. Some mammals possess a battery of glands. The ring-tailed lemur has glands on the genitalia, forearm and shoulder. The most common site for scent glands is the ano-genital region and the scent may be mixed in the faeces, as in the rabbit which deposits heaps of scent-marked droppings, or in the urine. The preputial (foreskin) gland of rodents discharges its secretions into the urine but in many other mammals the secretions are spread without the mediation of faeces or urine.

The glands, themselves, have been of interest to Man long before their value to the animals was appreciated. The fatty secretions were put to a variety of uses. That of the beaver, for instance, was used by the Ancient Greeks for the treatment of hysteria and heart palpitations and analysis has shown that beaver scent, or castoreum, contains salicylic acid, the active ingredient of aspirin. The anal gland secretion of the civet has been used for centuries as a base for perfumes. It has the property of 'fixing' or retaining volatile oils and releasing them slowly so that the perfume keeps its fragrance for a long time. The secretion is known as civet and the active principle, civetone, has been chemically isolated. Civet has the consistency of honey and is obtained from captive civets. Over the course of one week the African civet yields three to four grammes of civet secretion, which is ladled out every few days with a wooden spoon. The civet of the small Indian civet is used to flavour the local tobacco. The equivalent glands of the skunk and the stripe-necked mongoose produce the evil effluvium for deterring enemy attack.

Of a similar use to Man is secretion of the musk deer of eastern Asia which contains the active principle muscone. Musk deer have no antlers, as do most other deer, but they have long, sabre-like upper canine teeth. Both sexes have anal glands that give them a goaty odour but the male has a gland on the belly which secretes the jelly-like musk. When first extracted the musk is red but it becomes black and loses its obnoxious odour to reveal a pleasant scent when dry. Marco Polo noted the collection of musk along his road to Cathay. He stated that musk deer discharge their musk at the full moon and that they were so numerous that the air was filled with its scent at this season. Musk deer are still being hunted to this day, although their death is not necessary. The musk can be collected from the live animal by squeezing the gland. Apart from its perfumery use, musk has been used by the Chinese as a fertility drug, aphrodisiac and pain-killer and the Tibetans, incredibly, mix musk and tobacco with dung to make a snuff.

To spread the various message-carrying substances, mammals have evolved ritual patterns of scent-depositing or marking behaviour which may take the form of exaggerated movements of rubbing, chewing or eliminating. As such, they can be appreciated by the human observer who can record an animal's olfactory transmissions by visual observation of its behaviour without being aware of any scent. The dog's anointing of a lamp-post is the familiar example of scent-marking. It involves the leg-cocking movement and a high degree of control over the bladder muscles. Bitches, too, scent mark, by half squatting and half raising a leg. The hippopotamus spreads its dung by wagging its tail as it defaecates. The dung is splattered over nearby bushes, at a convenient nose height. Grey squirrels mark trees by chewing patches of bark and anointing them with urine. Scent-marking with the secretions of scent glands involves rubbing the gland against a suitable object. Mustelids, such as the badger, polecat and pine marten, drag their backsides against the ground to smear secretions of the anal gland. The civets back up to a suitable surface for the same purpose but some of the mongooses and genets rear up in a handstand to place the mark as high as possible.

If the gymnastic behaviour of the mongooses and genets is not quaint enough, bushbabies urinate on their hands, rub it on the soles of the feet and leave wet, smelly footprints as they move about. The tenrec licks an object to place its saliva then makes the mark more odorous by scratching itself and rubbing its foot in the saliva. Deer and antelopes need to be careful when scent-marking. The facial or preorbital gland lies just in front of the eye and was once thought to be an accessory smelling organ so that the deer could still scent danger when it was drinking. Its secretions are transferred to foliage or tree trunks by wiping it delicately to avoid the eye being damaged. Peccaries do not have this problem. Their scent gland is on the back and its secretion is smeared onto foliage as the animal pushes through the undergrowth.

This survey of scent-producing and marking is necessarily brief because mammals have almost as varied devices for transmitting olfactory signals as birds have plumages developed for visual displays and the two compete strongly for producing what are to our eyes, bizarre actions. Over the last thirty years careful study of birds has allowed ornithologists to decipher the messages that displays convey to other birds. It is now the turn of the mammals to come under scrutiny. They present the observer with the added difficulty that he cannot describe the scent as the ornithologist can record a bird display. An additional problem is that a single pheromone carrier may be used for more than one function. The urine of mice, as will be shown in the next chapter, can communicate fear, the sex of the mouse and its social status. It also mediates in aggression between males and in the sexual cycle of females. The scent of the young promotes a bond with the mother. It is not known whether the urine contains several odours producing different reactions or whether a single odour merely triggers the response that is suited to the particular situation.

# The social life of mammals

When the behaviour of mammals was first studied in detail it was found that many species indulged in scent-marking. Each individual transfers its odour to objects around the area in which it lives and neighbouring animals will investigate the traces of odour left behind. The first impression is that scent-marking is the equivalent of birdsong, that it serves to mark out a territory which is defended against rivals of the same species. To some extent this is true, but most mammals do not hold an individual territory with rigid boundaries against all comers as do many birds. The majority live in social groups, forming loose colonies or tight herds and packs, and, where there is a territory, it is usual for other animals to be allowed to trespass. A mammal is, nevertheless, concerned with demonstrating its presence, so that other mammals know when they are trespassing. It also has to attract a mate and, as a social animal, the mammal also has to show its relationship with other members of its group.

The original function of scent-marking seems to have been to ensure familiarity with an area. A mammal placed in a new cage spreads its own scent about to make itself at home, much as a child is comforted in a bedroom by familiar toys and pictures. For instance, R. F. Ewer's tame badger became agitated when brought into a strange environment and was calmed when presented with an object that it had previously marked. Among mammals living in communal burrows, mutual grooming and rubbing together imparts a colony odour, like that of the honey-bee, which allows recognition of group members. The marked area is the mammal's home but it is not necessarily a defended territory and it is now usual to talk of this area as the home range. This merely describes the area which the mammal covers in search of food and it may be not so much an area of ground as a network of paths. An otter has a home range of streams and pools with some interconnecting pathways. It marks these thoroughfares with piles of droppings on stones or wisps of grass to establish its familiarity and to notify its presence to other otters. Among domestic cats the home ranges of several individuals overlap. As each cat proceeds along its way, it deposits a series of scent-marks. Another cat coming across this trail will sniff the marks to determine their age. If old, it will proceed on its way; if new, it will turn aside to avoid an encounter with the preceding cat. In fact, if the first cat is aware that there are others near by, it will probably be more meticulous in its scent-marking because another function is to show an animal's intolerance of other animals. A scent-mark is a signal that the depositor may attack if disturbed. Thus, it is also a sign of social rank because higher-

*The gerbil is now a familiar household pet. The male marks its cage with the secretions of a gland on its belly. Any other male that is put in the cage knows that it is trespassing and tries to flee (Photo by Jane Burton)*

ranking animals are more intolerant of their fellows. This is particularly the case among mammals living in colonies.

Rabbits live in small colonies, each consisting of eight to ten individuals and occupying a territory based on a central warren of burrows with paths radiating through the grass and undergrowth. Within the colony there are two hierarchies, one of males and the other of females. The males are more intolerant of each other than are the females and there is considerable aggression between them. R. Mykytowycz, who studied the behaviour of rabbits kept in large enclosures, found that the dominant rabbits scent-marked more frequently than subordinates. Scent is deposited in two ways, as scented droppings deposited in piles around the territory and by 'chinning'. The droppings are scented with the secretions of the anal gland and chinning involves rubbing the secretions of a gland under the chin onto grass, tree trunks, old droppings and even the rabbit's mate and young. Mykytowycz found that the marked droppings of dominant males smelled more strongly than those of subordinates but it was otherwise difficult to determine when marked droppings were produced. On the other hand, chinning could be studied quantitatively and dominant males were found to chin more frequently than subordinates. This means that the dominant males are responsible for the defence of the territory. It was also found that dominant rabbits were heavier and that their chin and anal glands were proportionately larger. There is less of a difference in the chinning rates of dominant and subordinate females. They are more tolerant of members of their own sex than are male rabbits but they mark their nursery burrows with droppings.

Scent-marking by rabbits appears to have a two-fold function. It serves to mark out the territory, causing 'foreign' rabbits to turn back when they find scented droppings, and it demonstrates the rank of the marker. In hamsters and gerbils, dominant animals mark very much more frequently when placed in a subordinate's empty cage than when the procedure is reversed. The smell of another animal stimulates marking by a dominant animal but depresses the activity of a subordinate and smell alone may be sufficient to decide the outcome of an aggressive meeting. Ring-tail lemurs are primitive primates that live in Madagascar and have a complex series of olfactory signals. They have scent glands on the genitals, the forearm and shoulder and scent is either deposited on the branches of trees with the palms of the hands or is wafted towards the opponent by flicking the long, bottle-brush tail after it has been drawn over the glands. In this way two male lemurs have a 'stink-fight', presenting their scents to each other until the dominant lemur forces his opponent to withdraw.

Similar behaviour is seen in aggressive conflicts between deer, where scent secretions and urine are spread about by thrashing the vegetation with the antlers. It is a threat display similar to the display of the red breast by robins or the harsh long-call of herring gulls. For both mammals and birds the displays allow each male to establish its status without having to fight. The comparison of bird displays and songs and mammal pheromones can be extended because both have the second important function of bringing the sexes together. This does not seem to be import- ant in rabbits where the two sexes live a communal life in the warren but their relatives, the hares, lead solitary lives. Hares have comparatively small anal and chin glands but they have large groin glands, particularly in the female which appears to use them for attracting males from a distance.

On the face of it, the mammal's scents are acting as releaser phero- mones, altering the behaviour of an opponent or attracting a mate, but they also have a primer effect. The scent of the dominant animal appears to affect the physiology of subordinates, keeping them in a state of retarded development, rather as queen substance acts on honey-bee workers to prevent the maturation of their sex organs. The primer effect of mammalian pheronomes has been closely studied in the rats and mice. They have been found to have a profound influence on breeding physiology and behaviour and to play an important part in the regulation of the numbers of animals in a population. The means by which animal populations are kept in check – the factors involved and the manner in which they operate – forms one of the central problems of zoology. Population growth must be prevented from continuing to a level where the animals become overcrowded and the food supply is eaten out. There is increasing evidence that, for some mammals at least, the factors regulating population growth operate through the medium of phero- mones that control reproduction and that the pheromones, external chemical messengers, are governed by hormones, the internal chemical messengers.

In mice, and many other rodents, the main sources of pheromones involved in reproduction are the urine and the secretions of preputial glands. As with the chin and anal glands of rabbits, the size of the preputial glands of rodents suddenly increases as the animal becomes sexually mature at puberty and the glands are larger in males than in females, where they occur as clitoral glands. Because the preputial-gland secretions are discharged into the urine, it is not always easy to separate their effect from that of the urine itself but, between them, urine and preputial-gland secretions act as releaser pheromones causing fighting between males and attraction of females to the males. Female mice are not attracted by male pheromones when they are pregnant. Their reaction appears to be blocked by the high level of the hormone progesterone which circulates in the blood of pregnant mammals. Progesterone is primarily concerned with the maintenance of pregnancy and the accompanying physiological changes and its use in 'switching off' courtship behaviour is a new and interesting discovery.

Progesterone does not block the female mouse's ability to smell the scent of male mice. Only her behaviour towards them is changed and male scent continues as a primer pheromone affecting the physiology of the pregnant female. This effect was discovered by accident, as often happens in scientific investigation. Dr Hilda Bruce happened to place a newly-mated female mouse with a male that had not been her partner in mating. The strange male mated with the female, which was unexpected because female mice were expected to reject the advances of the male once they had been mated. On investigation, Dr Bruce found that the strange male blocked the pregnancy, providing that it was introduced to the female not more than a week after she conceived. Blocking is most effective if the introduction of the male is made within four days of mating and is caused by the developing embryos being prevented from implanting in the wall of the uterus. When this happens, the pregnancy fails and the mouse comes into oestrus (on heat) again. The level of progesterone in the blood drops and the female again becomes susceptible to the courtship of male mice. Once implantation has taken place, however, the strange male has no effect on the pregnancy and is rejected by the female.

Pregnancy blocking still occurs if the strange male mouse is no more than caged within the female's run or if his urine drips into it and he has no physical contact with the female herself; and implantation of the embryos even fails if the female is placed in a box where a strange male has been kept previously. Hence, it can be concluded that it is the odour of the male that affects the pregnancy. What the odour is and how it acts is not known but a female mouse must know the scent of her first mate so that she can distinguish other males as strangers. Why a strange male should cause what is now known as the Bruce effect is not altogether clear. There is no obvious advantage to the female but it will increase the number of offspring to which the strange male contributes. Natural selection takes place through the 'fittest' individuals of a species producing the most offspring and the Bruce effect could be a mechanism by which a male mouse can increase his progeny at the expense of other males. In the wild, where a male mouse lives with one or more females in a defended territory, it may be that a male which displaces another male from its territory is 'fitter' than the ousted male and it is advantageous to the species for him to produce offspring as soon as possible. By the Bruce effect, his progeny immediately usurp the place within the female's uterus of those of the ousted male.

Female mice also produce a pheromone or pheromones (the chemical identification of rodent pheromones has not been completed). The Bruce effect is reduced if several females are exposed to the strange male at the same time, presumably because each female is receiving a female pheromone which counteracts the effect of the male pheromone and prevents her from losing her litter. Moreover, when four or more female mice are housed together in the absence of any male, their sexual cycles cease. They no longer come on heat and they are said to be anoestrous. This is known as the Lee-Boot effect after the Dutch discoverers van der Lee and Boot. If a male is then introduced to the group, the females' cycles start simultaneously and they come on heat three to four days later, a phenomenon called the Whitten effect.

The role of the Lee-Boot and Whitten effects in the lives of wild mice is not known but there may be some advantage to the mice in having synchronised mating and pregnancies. The urine of female mice is known to inhibit the aggressiveness of the male, which is helpful in allowing mating to take place. There is a possibility that the females are most effective at checking the male's aggression and inducing courtship when they are on heat and that the effect on the male is increased when several females come on heat together.

There is one further primer effect known for mouse pheromones. Mice living in crowded conditions develop symptoms of stress and a mouse kept in solitary confinement develops the same symptoms when the odour from a crowded cageful of mice is blown through its box. The symptoms include swelling of the adrenal glands, disturbance of the animal's behaviour and physiology and, of greatest importance, the ability to reproduce successfully becomes impaired. The mice are less likely to mate properly, litters are lost during pregnancy and maternal behaviour towards the baby mice is disturbed. All these symptoms are signs of the stress syndrome, a condition that is often seen in animals living in the overcrowded conditions of high population density. That it can be induced by odour alone rather than by physical overcrowding is crucial to understanding the basic function of the pheromones that have been described. Some zoologists think that animal populations, particularly those of rodents which show regular cycles of abundance, are regulated by means of stress. When overcrowded, their adrenal glands are found to be excessively large, they are aggressive and their breeding fails. The Bruce, Lee-Boot and Whitten effects and a 'stress pheromone' are the first evidence we have of signals between mammals which can provide a means of controlling population growth. They will promote breeding at low-population densities and inhibit breeding at high densities.

Consider a population of mice living at a reasonable density. Each male holds a territory and serves a small number of females. Both sexes are familiar with the odours of other animals in the territory and the male repulses intruding males, when they cross the boundary into the territory, by fighting or scent marks. The females bear his offspring and nurture them until they become independent. For the young animals the first weeks of independence are crucial. Their pheromone-producing glands are small so they are of low social rank and will be chivvied and chased by their superiors. The chances of the young animals subsequently obtaining a place in the society of their species depends on the density of the population. If numbers are small, they can find territories of their own but, as numbers increase, they will become difficult to accommodate. They will become outcasts, access to food will be difficult and stress develops through continuous aggressive encounters. Eventually the young unestablished animals die. Furthermore, as numbers in the population pass the optimum level that the environment can support, as they may do when food has been particularly abundant, life for even the established animals becomes difficult. The territories become squashed together and there are more boundary conflicts. Even within one territory, the members come into conflict. Symptoms of stress appear and breeding fails, through the physical effects of over-crowding and the stress pheromone. There is sometimes a cataclysmic crash in numbers, aided by starvation and the predators drawn to a good supply of food. But, after a period with numbers at a minimum, the sur-vivors eventually rally to restore their losses. Pheromones advertising the position of a mate and his or her readiness to mate then help to bring scattered animals together and the Whitten effect accelerates the sexual cycle of females to bring them on heat when they meet a male.

In this cycle of numbers linked with social behaviour there is an interaction of hormones and pheromones. Young males have small pheromone-producing glands and cannot compete in 'scent-fights' with dominant animals. The young male animal is kept in its subordinate place by encounters with dominant males. Their attacks cause it stress which stimulates the adrenal glands to produce hormones that keep the phero-mone glands and testes small. When the dominant animals die, the source of stress is removed, the pheromone glands and the testes start to func-tion properly and the young animal achieves a position of dominance. Its pheromones are then capable of affecting the behaviour of other males and influencing the sexual cycles of the females. Somehow, they affect the pituitary gland of the females, which secretes the hormones that control the physiology and behaviour of reproduction.

So here is evidence for a mechanism by which the lives of mice and presumably other small mammals are regulated. It is a language which is simple but which conveys sufficient information for the animals to adapt to variable conditions. The realisation that it is through the language of smell that the lives of these mammals are co-ordinated has revolu-tionised our ability to understand the pattern of these lives. It was once a mystery how the life in a colony of rabbits or mice was controlled and how mating and parental behaviour were ordered. All that could be said was that some sort of instinctive processes must play a part, that the mammals were born with an ability to live together and rear their young. Now we can see the link which integrates the behaviour of one mammal with that of its fellows. It is the communication of the physiological state and social rank of that animal to the others.

To date, it is only in the rabbit and house mouse that the use of pheromones has been studied in depth. We can guess that lives of other mammals will also be found to have a complex language of smell. There is the ring-tail lemur with its variety of scent glands. It is one of the more primitive primates, a group whose more advanced members have come to rely especially on vision and hearing for communication. In monkeys and apes, status, sexual condition and other social characters are indicated by calls, like the territory-advertising hoots of the howler monkeys, or by visual signals, such as the obscene-looking swellings and colourings of many monkeys. Scents are less obvious in the monkey world and have almost disappeared in apes and Man, yet the use of pheromones has been retained apparently to signal the physiological state of the animal. Pheromones are convenient signals as the hormones that control the physiological state of the animal can simply trigger the secretions of the pheromone glands. No elaborate behaviour or bodily change is needed, as there is for auditory and visual signals.

A conspicuous feature of the sexual cycle of female monkeys and apes is the colouring or swelling of the genital and anal regions, when the animal is on heat. The female rhesus monkey displays a red patch of skin which informs males of her condition but she also secretes a pheromone from the vagina. The pheromone renders the female attractive and leads males to mate with her. It has been aptly named copulin and is, presumably, secreted in response to the hormonal changes that take place at the time of ovulation but it is not yet known to what extent the secretion coincides with the red colouring. They may both signal the same message, which would appear to be pointless, or the message of each may differ slightly. The colouring could generally indicate the female's condition while the pheromone may serve to focus the male's attention and trigger mating. There is some support for this idea from an experiment in which the genital regions of neutered females were caused to turn red by applying female hormones. The males took little notice of them but, when the redness died away and copulin was applied, the males became very interested and mating took place.

It is most likely that there are other pheromones to be discovered among the monkeys and apes. Copulin-like pheromones have already been discovered in species other than rhesus monkeys. Adult male chimpanzees and gorillas sometimes emit a strong smell when excited but it is not known whether this is a pheromone and conveys any message. It is not likely, however, that any pheromone system as complex as that of the lemurs will be found. The higher primates are losing the sense of smell in favour of eyesight and hearing. The process has continued in the highest primate, Man, where eyesight is indisputably the most important sense. But we have now found that the sense of smell is used for communication throughout the animal kingdom and in animals where its use had not previously been suspected, so could it be that we have our own pheromones that have been passing unnoticed literally under our noses? They need not be very important, merely relics passed down from our lemur-like ancestors but there is now more than a suspicion that they exist. If they do, they are of more than passing interest. Pheromones, we have seen, are important in the interactions of social animals. Man is the most social of animals and we might understand interactions between people better if we find that we have overlooked a complete system of communication between them.

# What about Man?

The survey of the animal kingdom has shown the extent to which the sense of smell is used by a wide variety of animals. It is only in the last decade or so that this has come to be appreciated. Previously we could only hint at the use of smell by many animals. Some were not known to possess a sense of smell and for others its role in their lives could only be guessed. Organs whose function was unknown have now been found to be scent glands, for instance the preputial glands of rodents. The question now is whether we can extend these findings to include Man. Ever since Darwin showed that Man was related to the apes and so to the rest of the animal kingdom, it has been the custom to apply the techniques and theories of zoology to the study of Man himself. We now look at Man as a special sort of animal who must still obey the basic laws of nature if he is to survive.

In the nineteenth century, academic zoology was largely a matter of comparative anatomy in which the study of bones was used to demonstrate the relationships between animals. When applied to the human skeleton the study showed clearly both our relationships with monkeys and apes and our own special features – the upright gait for walking and the opposable thumb for gripping, for instance. From the basis of comparative anatomy, the study of zoology has broadened and we are now entering a phase of the close study of the social life of animals, which includes communication. Instances of such work have been given throughout this book, from Fraser Darling's pioneer study of red deer to observations on the communities of hunting dogs and rabbits and the organisation of beehives. Conclusions from the study of animal behaviour are now being applied to the study of human life. Desmond Morris's *The Naked Ape* and Robert Ardrey's *The Territorial Imperative* seek to explain our sexual and aggressive behaviour, by relating it to conclusions drawn from animal behaviour, and thus to establish the roots of our behaviour in the same way as the anatomists traced the origins of our physical appearance. If our fundamental nature can be discovered, so it is argued, we shall be better equipped to understand our emotions and control lapses from civilised behaviour. The last chapter showed how social rank and sexual behaviour of mammals are often mediated by smell. So, following the trend of treating Man as an animal, it is worth looking for evidence that Man, himself a mammal, is also using his sense of smell in a social context. Already there is some confirmation that Man does have a language of smell. The evidence is slight but sufficient for speculation to be made about its use.

Before discussing the use to which Man puts his sense of smell it would be well to determine just how sensitive is this much maligned sense. From anatomical and behavioural considerations it is clear that Man's sense of smell does not measure up to that of many animals, such as the domestic dog, but our noses are still extraordinarily sensitive. One olfactory receptor cell in the human nose requires only eight odorous molecules for stimulation and forty or more need to be stimulated for a conscious sensation of a smell. Sensitivity to such a small number of molecules explains how a man can detect the odour of a footprint on blotting paper and why one milligram of skatole (the odour of faeces) can render unpleasant the atmosphere in a hall 500 metres long, 100 metres wide and 50 metres high. The human nose can also distinguish a huge variety of odours. A skilled perfumier can distinguish up to 10,000 odours. He will have a range of 2,000 ingredients for use in his work and can readily identify half of these.

The perfumier trains his sense of smell. He is better at smelling than the rest of us because he spends his day working with scents and, like any other expert, his prowess is developed by practice. The birdwatcher identifies, with no trouble, a species that to the uninitiated, if he sees it at all, is no more than a little brown blur disappearing into the foliage. The birdwatcher has spent many hours learning the distinguishing features of birds and is always on the lookout for them. So with odours, most of us do not notice them, let alone bother to learn them, but the perfumier studies them and is always conscious of them. But training is not every-thing; some people are insensitive to smells as the colour-blind person is to colours. There is also a variation in sensitivity which is related to the anatomy of the nose. In some people, the shape of the nasal passages hinders the passage of odour-laden air into the nasal clefts so that they may be unaware of odours unless they seek them by sniffing hard. But the most important reason for our lack of appreciation of odours is probably that our noses are in the wrong place. The perfumier can lift his sample bottles to his nose but we generally do not go around sniffing like a dog. We do not notice the smell of a bare footprint because our noses are usually 1·5 metres above the ground. The evolution of an upright posture has taken our noses away from the source of many odours. Shrews, mice, cats and dogs walk with their noses near the ground and one can only enter their world by descending to hands and knees, as will be appreciated by a gardener grubbing weeds from a border or a naturalist searching for specimens among the grass stems. Apart from a powerful background odour, the odours of individual plants and animals can be appreciated and sometimes used for identifi-cation.

It would seem, then, that Man has quite a good nose but that during the course of evolution he has left the world of smells. The first primates were ground-living, rat-like animals. Their sense of smell was extremely important but as later primates took to living in the trees the sense of smell has steadily atrophied. Nevertheless, pheromones still play a part in the lives of monkeys although the sense of smell is not one of their main senses. Is there, then, still a use for pheromones in the lives of Man?

Taking first behaviour used in the establishment of a territory or range and dominance over other animals, there are very few indications of pheromones being used in relationships between men. It has been said that Australian aborigines mark territories with urine but this has only been recorded as an anecdote and is hardly the basis for further specula-tion. Among a certain tribe in New Guinea it is customary for friends to exchange odours when parting. A hand is pushed under the armpit, smelled and rubbed over the body. The gesture may be just a ritual but a ritual is the stilted performance of a once meaningful action – like the friendly salute which once showed that weapons were not held. The New Guinea ritual shows amicable intent by a gesture but the body odour may confirm the friendly emotion. There are also stories of the emotion of fear being communicated by smell. Some people claim that they can smell fear in others and witch-doctors are said to detect miscreants by this ability. People suspected of a crime are lined up and treated to a performance of 'magic' in which the witch-doctor appears to be divining the identity of the guilty person. In practice, the latter, providing he believes in the efficacy of the 'magic', is frightened by the rigmarole and gives himself away by his smell, even if remaining outwardly composed.

In Western civilisations the ability to smell emotion has been lost although there is a hint in the German *mann kann jemanden nicht riechen*, literally 'I cannot stand his odour' and equivalent to the English expression 'I cannot stand the sight of him'. The English idiom is the most logical as sight is our main sense but maybe the German harks at a past use of smell. It is even possible that the Germans are right and that we do react to an individual's odour but only at an unconscious level.

Yet this is all anecdote and surmise. There is no tangible proof of odours conveying aggression, fear or friendship, but there are slender and intriguing grounds for thinking that odours act as messengers in sexual relationships. Artificial odours or perfumes have had a long use as aphrodisiacs but they function more by association than by any pheromonal effect on the receiving individual. The true link between smell and sex lies in the simple and well-known observation that women are better than men at smelling certain substances. At one time this was put down to women's relative abstinence from tobacco and alcohol. Such an explanation is less tenable today, but careful experiments do show that women detect a range of substances at far lower concentrations than can men.

In 1948 L. LeMagnen, a French physiologist, discovered by accident that adult women were very sensitive to the smell of exaltolide, whereas

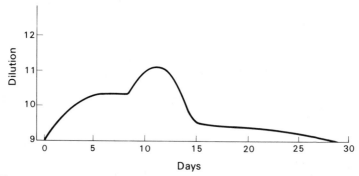

The variation in sensitivity to the smell of exaltolide through the menstrual cycle. Women perceive the smell of the most dilute solutions of exaltolide at ovulation, usually between the twelfth and fifteenth days of the cycle.

man or young boys and girls either failed to smell it or could do so only feebly. A woman's sensitivity to exaltolide, and many other chemicals, varies through the course of her menstrual cycle. She is most sensitive at the time of ovulation and is relatively insensitive during the first months of pregnancy. Exaltolide is extracted from angelica, the root of which is used to flavour liqueurs such as Benedictine. It smells like, and is chemically similar to, civetone and muscone, the pheromones used by male civets and musk deer respectively. It is mainly the musky odours to which women are more sensitive than men. 'Boar-taint' of pork is another musk-like substance smelt more readily by women. It appears to be made from a musky substance secreted in the urine of boars and, as the name suggests, it imparts an unpleasant odour to pork made from uncastrated boars. The musky substance is a pheromone which causes a sow on heat to stand in the arched mating position when a boar approaches. A similar substance is found in human urine.

A pattern can now be seen emerging. Women are sensitive to substances that have a musky odour, that are used as pheromones by

*Structural formulae of musk-like compounds and the male sex hormone, testosterone.*

various mammals and are often associated with the male animal. Moreover, these 'musks' are chemically related to the male hormone, testosterone. Whether there is a cyclical sensitivity to all these substances, as to exaltolide, is not known. It would also be interesting to know whether contraceptive pills that abolish ovulation also abolish sensitivity to exaltolide. So is there any significance in women being sensitive to what appear to be male scents and perhaps being hypersensitive to them during the period of ovulation?

Although women have a cycle of fertility, the menstrual cycle, there is no fixed period for mating as in the 'heat' period of other mammals when the female is receptive at ovulation. Neither do men have a cycle of potency like many male mammals. In short, the human species does not have a mating season and there is no great need for a sex pheromone, like copulin in monkeys, or other specific signal to bring the sexes together and promote mating at the correct time. There would be distinct draw-backs in the modern way of life if there were such a pheromone, because female monkeys put 'on the pill' are ignored by the males. Contraceptive pills prevent them coming on heat, no copulin is produced and the males are not aroused. The human habit of mating throughout the fertility cycle is unusual as is the resulting production of young at any time of the year. For primitive Man there may have been an advantage in having a mating season so that children were born during the season of abundant food, but after our ancestors had learned to store food and build shelters, this would be of less importance. On the other hand, an important step forward in the evolution of Man was the addition of the male to the basic mother-young family unit of other primates. Continuous mating would, and does, help maintain the bond between parents in a species where the co-operation of the male is needed to assist rearing slow-growing offspring.

Although men are not attracted to women solely during the fertile periods, there is, perhaps, some sense, biologically, in women being sensitive to a male pheromone during this period. For the purposes of reproduction, it is advantageous to ensure she is particularly receptive to advances when able to conceive. With the modern trend to small, or non-existent families, this is more of a disadvantage, but throughout the animal kingdom the first aim of a species is to produce as many offspring as possible. It is only recently that Man has sought to change this biological ideal, a change that is by no means sought by all people.

Whether men do secrete a pheromone for the purpose of stimulating fertile women is, once again, a matter for conjecture. In other mammals, pheromones are produced in specialised scent glands, or in faeces, urine, saliva and sweat. Of these possibilities, sweat and urine are the most likely vehicles for olfactory signals in Man. Human urine contains a musky substance which is significant in the light of LeMagnen's findings but there is no evidence for the use of urine in human social behaviour. Sweat is a more likely carrier of pheromones. Sweat glands are found all over the body and sweat carries the characteristic body odour. To some extent the body odour (excluding breath odour) is related to diet. The odour of Eskimos is due to a diet of fish; Mediterranean people smell of garlic and onions and northern Europeans have a cheesy, buttery smell. Within these broad categories there is much variation and people with a good sense of smell can identify individuals by odour alone, using true body odour and not tobacco, perfume or other acquired odours.

The main sites of sweat secretion and body odour are the armpits and the groin, which also bear the main body hair. In Man's upright posture, these areas get hot and sweaty and the hair increases the surface area from which sweat can evaporate and so promote cooling. As such, the body hair could be acting as a pheromone disseminator like the hairpencils of a butterfly or the fluffy tail of a lemur. The armpits are best placed as disseminators of pheromones because they are on almost the same level as the nose. Confirmation that the armpits are a source of pheromones lies in the gesture of the New Guinea tribesmen at parting and in the oft-quoted anecdote from *Psychopathia sexualis* by Krafft-Ebing. A young man was reputed to have great success with girls. After a dance, he would wipe the perspiring brow of his partner with a handkerchief that had been carried in his armpit. Apparently his body odour acted as an aphrodisiac and the young man claimed that his technique was highly successful. There appears to have been no attempt to substantiate this interesting technique although it opposes the usual statement that there is no such thing as a real aphrodisiac and although it is the only evidence for a male-sex pheromone in Man.

One other human pheromone has come to light. Its existence has been sufficiently well investigated to be reported in the top scientific journal *Nature*. A study of students living in a dormitory of an American women's college showed that the menstrual cycles of close friends and roommates became synchronised. The effect was due solely to the time spent together and was not caused by similar patterns of time spent awake or asleep, to similar eating habits or to age. The only conclusion is that there seemed to be some unconscious communication about physiological status between women spending much of their time together. As the menstrual cycle is controlled by hormones, there could possibly be a hormone-linked production of a pheromone, as is known to occur in some mammals. Direct, verbal communication between the women was ruled out because few of the students were aware of the timing of their friends' cycles.

The study also showed that the presence of males affected the length of the cycle. Women meeting men less than three times a week had cycles on average of thirty days whereas those that saw men more often averaged twenty-eight and a half days and individuals reported that their cycles became shorter and more regular when exposure to males increased. Unfortunately the report does not say whether the 'exposure' involved physical contact, which may range from ballroom dancing to sharing a bed, but the wording suggests that mere presence of men was the stimulus. It would also be useful to know whether the exposure was of a social nature or was merely routine contact during everyday activities. If the latter, there is a greater likelihood of a pheromonal stimulus rather than the psychological stimulus of being in the company of eligible or potential mates. There is, of course, no concrete evidence for a pheromone from males being involved at all but the situation does recall the Whitten effect of male mice initiating the oestrus cycles of a group of female mice.

The evidence for human pheromones available to date is tantalising rather than convincing. There are only some anecdotes and a couple of scientific investigations to show modifications of behaviour or physiological state could involve the sense of smell. No attempt has yet been made to isolate human pheromones and we can do no more than speculate about their possible function in human behaviour. It is not likely that they are of great importance. Susceptibility to male odour at ovulation may have been useful to primitive Man or may be a relic left from prehuman ancestors but there does not seem to be any value in synchrony of female cycles in a cloistered life. Human behaviour is not triggered by simple stimuli, as is the courtship of moths for instance, but by a complex of sensory and psychological stimuli in which memory and association play an important part. The absence of a single stimulus does not greatly affect behaviour and that this is the case in the human sense of smell is shown by the increasing use of deodorants and perfumes. They are designed for the removal and masking of precisely those body odours that are likely to be pheromones. As Alison Prince wrote in the *Observer* magazine, the use of these products is 'simply a form of cutting off the nose to spite the face'. It would be too much to say that the use of deodorants disturbs our behaviour by removing a form of communication, as it is obvious that many users find no loss to their social life. Human pheromones may not be important but it would be a pity to abolish them. Plain food may be edible, but sauces make eating a pleasure and the same holds for other basic human activities.

# Further reading

Butler, Colin G., *The World of the Honeybee*, London, Collins, 1975.
Ewer, R. F., *Ethology of Mammals*, Plainfield, New Jersey, Logos 1968.
Free, John B. and Butler, Colin G., *Bumblebees*, London, Collins, 1959.
Sudd, J. H., *An Introduction to the Behaviour of Ants*, London, Arnold, 1967
Wright, R. H., *The Science of Smell*, London, Allen & Unwin, 1964.

# Index

# Index